3rd Edition

The
Backyard Beekeeper

An Absolute Beginner's Guide to Keeping Bees in Your Yard and Garden

The
Backyard Beekeeper

3rd Edition

An Absolute Beginner's Guide to Keeping Bees in Your Yard and Garden

Quarry Books
100 Cummings Center, Suite 406L
Beverly, MA 01915

quarrybooks.com • craftside.typepad.com

First published in the United States of America in 2014 by
Quarry Books, a member of
Quarto Publishing Group USA Inc.
100 Cummings Center
Suite 406-L
Beverly, Massachusetts 01915-6101
Telephone: (978) 282-9590
Fax: (978) 283-2742
www.quarrybooks.com
Visit www.QuarrySPOON.com and help us celebrate food and culture one spoonful
at a time!

10 9 8 7 6 5 4 3 2

ISBN: 978-1-59253-919-2

Digital edition published in 2014
eISBN: 978-1-62788-039-8
First Edition published in 2005
Second Edition published in 2010

Library of Congress Cataloging-in-Publication Data available

Design: tabula rasa
Page layout: Sporto
Illustrations by Michael G. Yatcko, 18; 56; 64; 72; 77; 78; 93
Cover images: Getty images (main image), © Valery Rizzo/Alamy, (bottom, left);
Shutterstock, (top, middle & left)

Printed in China

Dedication

This book, the process that brought it to be, and the evolution of the information provided here is hereby dedicated to Professor Chuck Koval, Extension Entomologist, University of Wisconsin, Madison—who first let me in and showed me his way of sharing information. I miss his good advice and his humor, but not so much his liver and onions.

To Professor Eric Erickson, USDA Honey Bee Lab, Madison, Wisconsin, then Tucson, Arizona, and now retired, who made me learn about bees, and who encouraged me to learn, and to use what I learned to help those who could use that information.

To John Root, President (now retired), of the A. I. Root Company, Medina, Ohio—who hired me to shepherd his magazine, *Bee Culture*, and who let me bring together all that I had to take his magazine to the next generation of beekeepers.

And of course to Kathy, who has put up with all of this for all of these years, and enjoys it as much as I do.

And last, first, and always . . . the bees. Where would we be without them?

∗→ CONTENTS ←∗

Preface

Since the first edition of this book was published, a tsunami of changes has crashed over the beekeeping world. It seems almost too many to number, though I will try because it is important to delineate why this book has been updated and revised, now a second time. Though much has changed, much has stayed the same. I have retained the sections and the information that have not changed. But the ideas, techniques, and principles that are no longer viable are no longer here.

Of course much of the attention bees, beekeepers, and beekeeping have received in the past several years can be attributed to, as the scientists will tell you, the multi-factorial causes identified that will produce an unhealthy, or dead, colony.

In spite of all the attention, research, money, press coverage, and the discoveries that *weren't* the solution to what was being called Colony Collapse Disorder, the final answer still remains elusive. And, in fact, it turns out no single solution is the answer, but rather a whole slew of issues coming together in a beehive. But along the way many serendipitous discoveries were made. For instance, honey bees are increasingly exposed to a witch's brew of sly new crop pesticides that were definitely poorly tested and poorly regulated before being released. In addition, several years of drought and poor foraging have reduced both the quality and quantity of available food for bees almost everywhere. This coupled with an increasing diet of monoculture crop monotony has created significant nutritional distress.

The bane of beekeepers worldwide continues to be *Varroa* mites. Bees resistant to or tolerant of *Varroa* mites remain more wishful thinking than actual fact, with the exception of the Russian strain of bees. But these, too, are difficult to obtain, though they are finally becoming somewhat available. However, it is the cursed viruses that *Varroa* has unleashed that are at the heart of most of these colony deaths.

Viruses affecting honey bees are present in all colonies, but they remain dormant until a honey bee is attacked and wounded by a *Varroa* mite. This injury is bad enough, but a honey bee's immune system is challenged by this attack, and the viruses are able to become active. They are then spread from bee to bee, from queen to egg, from bee to queen. Vertical and lateral transmission is common, and once a colony is infested, most bees are doomed.

The stress on some colonies from moving from place to place has not let up, while *Nosema ceranae* remains a sometimes deadly, but elusive, parasite.

The symptoms of Colony Collapse Disorder went mostly unnoticed in the beginning, then made lots of noise for several years, and were studied to death. Finally the world agreed it was not a disease or disorder, but the effects of some combination of nutrition, pesticides (both beekeeper applied and industrial agriculture), *Varroa*, viruses, *Nosema*, stress, and probably other factors. The disorder has been studied from every angle, every perspective, and every scientific process—we know more and more about the problems we face yet have done virtually nothing to solve them.

During the ensuing years roughly 30 percent of the bees in the United States perished each winter, sometimes more, seldom less. The situation has been similar in most places around the world that keep bees. Countries have banned some classes of pesticides in an effort to stem the tide; dietary supplements for bees have flourished; new *Varroa* treatment chemicals have come forth, and vanished; and a whole new generation of beekeepers has appeared, concerned and committed to helping honey bees.

And this is why I have revised this book. We have learned good lessons: In killing *Varroa* we have sullied our hives and rendered them nearly unlivable for our bees, and in the process dirtied every bit of the beeswax in this country. We know more about what bees need to eat and when they need it, and we know we must feed them because nature no longer can. We know *Varroa*, kind of, and what it can do, and we keep trying to stop the slaughter—and we have in some cases, but not in most. We know that the world of industrial agriculture keeps getting better and better at killing insects, and our bees are insects. We know that subtle doses of these killing sprays may be worse than a fast and clean death. These chemicals linger and maim and slowly destroy, but remain hidden so we don't know what the problem is. But we've learned ways to counteract these problems, to fix what went wrong, and to do things a better way—so *Varroa*, not enough food, pesticides, and viruses don't stop us from enjoying the ways of our bees.

You will begin your beekeeping adventure well-armed with all this new information plus the tried and true ways that remain. Add to this that urban and rooftop beekeeping has risen and spread like warm honey on a hot biscuit. If you are part of this movement, then what's inside will be a welcome addition to your citified beekeeping endeavors. You are, right now, light years ahead of where beekeepers were even five years ago. You will be a smarter, better beekeeper.

Backyards are good places to keep bees because they are close; urban areas support bees well with diverse and abundant natural resources; and bees are the pollinators of choice for gardens and landscape plants all over the neighborhood.

With this book, a bit of outdoor wisdom, and a colony or two of honey bees you will truly enjoy the art, the science, and the adventure of beekeeping. You will enjoy the garden crops you harvest, the honey you and your bees produce, and the beneficial products made from the efforts of your bees and your work.

What could be sweeter?

Enjoy the bees!

—Kim Flottum

✦ INTRODUCTION ✦

I've been around bees, beekeeping, and beekeepers for more than forty years, and in spite of the day-to-day issues we face, there has never been a better time to have a few honey bee colonies in your backyard. Honey bees pollinate the vegetables in your garden and increase the production of your orchard trees. In fact, the crops that are pollinated in part or in whole by honey bees supply us with an incredible amount of our daily sustenance. Scientists and crop producers tell us that honey bee–pollinated plants may account for a third or more of our daily diet. With fewer honey bees what we could eat is diminished, and the mundane grasses—wheat, rice, corn, oats, and barley—would take on an even more important role in what we routinely consume.

Perhaps as importantly, honey bees make wild plants more productive, more luxurious, and more nutritious for the wildlife that relies on them for sustenance. Many birds, rodents, insects, and other animals would be foodless without the myriad weeds and wildflowers that grow wherever we let them.

Honey bees, beekeeping, and beekeepers have received a startling amount of attention due to two facts: the dramatic decline of the honey bee population, and thus the threat of losing their bounty and benefits; and the sudden awareness of both the loss, and what could be done to reduce its impact and even enhance the quality of life of the honey bee population.

Already aware of the blessings of having honey bees in their lives, suburban, urban, and city gardeners and growers capitalized on these events and made sweeping changes in zoning and livestock laws in many, if not most, major cities. Now, along with chickens and other small stock, honey bees are back in the neighborhood—pollinating garden crops and fruit trees, street trees, and window boxes. Bees, beekeepers, and beekeeping—they're all back where they should have always been.

With garden harvests a part of your life, cooking up simple dishes using your bounty is probably already second nature. Adding bees to your routine and adding your own honey to the table will allow you to reap what you sow all year long.

But where do you start? What do you need? And, most important, how much time will it take?

If you're like me—and most other people today—time is important. So, how much time does it take to set up and take care of a couple of colonies of bees? Tending bees is a lot like taking care of a garden. There's a flurry of activity in spring, maintenance in summer, and harvest in fall. Over a season, your bees will take a bit more of your time than

🐝 *A rare view of part of the wood factory at the A. I. Root Company, before 1900*

you spend caring for your cat, but less time than you spend with your dog. Like any new activity, there's a learning curve in beekeeping, so the first season or two will require more of your attention than will be needed once you have some experience under your belt. And like a garden, there's prep work before you begin each new season and some equipment you'll need to get started.

Oh, and bees do sting. Let's get this right out front: They aren't out to get you, but they will protect themselves when disturbed. But think for a moment—bramble thorns scratch, mosquitoes bite, and yellow jackets are just plain nasty. Cats and dogs also scratch and bite; it's the way things are, plain and simple. But you wear gloves to prune your rose bushes, you wear mosquito repellant when outside at dusk, and if you don't tease your pets, they probably don't give you too much grief. It's the same with your bees. Work with them, use the tools you have for good management, and wear the right gear. Even when using gloves and long sleeves, stings happen, but if you are smart and prepared, they will be rare events. When brambles, mosquitoes, cats, dogs, the scratchy stems of zucchini plants, or honey bees cause that momentary ouch,

REV. LANGSTROTH VISITS THE A.I.ROOT COMPANY

L.L. Langstroth, holding the frame he invented, sitting in the beeyard at the A. I. Root Company, in Medina, Ohio, many years after his "Eureka!" moment

figure out what you did to cause the ouch, utter a soft curse, rub the spot, and move on.

So, if having a couple colonies of honey bees out back sounds like a good idea because you want a better garden, more fruit, honey in the kitchen, maybe some beeswax candles, skin creams, and other cosmetics for the bath, let's find out what thousands of beekeepers already know.

In the Beginning

Honey has been a source of sweetness since people and bees first met. Initially, the only way to gather honey was to rob it from bees nesting in caves or trees. This was hard on the bees. (See "But First, a Little History," page 13.) This evolved to people keeping bees in baskets, but to harvest the honey the bees were killed with sulfur fumes. Along came gums, whole sections of trees the bees lived in, removed to a suitable location for their keepers. The honey could be removed without killing the bees, but it was still destructive and the bees had to rebuild much of their nest every year. Next came boxes, and though they were easier to move, they still had the same destructive qualities. Honey could be harvested, but to

do so meant the beekeeper had to remove and destroy many of the beeswax combs filled with honey.

The way we keep bees today can be traced back to the introduction of the modern hive in the mid-1800s. L.L. Langstroth, a minister who suffered from a nervous condition now thought to be a bipolar disorder, became a beekeeper to ease his discomfort. He became wise in the ways that bees were being kept in all parts of the world and experimented with his own bees, looking for a way to keep bees from fastening their combs to the top and sides of the boxes and gluing all parts of the hive parts together with propolis.

Beekeepers in other parts of the world had already discovered top bars and frames, therefore, most of the comb wasn't attached to the top and sides of the box. They still, however, had trouble with propolis, especially when used to fasten the top bar to the underside of the cover.

The story goes that one day as Langstroth was walking home from a visit to a beeyard, he had a vision of a frame—a complete square of wooden strips that surrounded and contained the beeswax comb. The bees could attach their comb to the strips rather than the top and sides of the hive box. He saw a means of "hanging" the frame inside the box the bees lived in. This kept the comb separated from the top, sides, and bottom of the box by just enough space for the bees to pass. This space, between ¼" and ⅜" (6 mm and 1 cm) came to be known as the *bee space*. This concept revolutionized beekeeping and is the perfect example of what working *with* the bees can accomplish, rather than working *against* them. This fundamental design has remained virtually unchanged since Langstroth's discovery.

Langstroth's recurring mental health issues and his preoccupation with patents and the rights to manufacture his hive slowed down acceptance of this design, though it was adopted by some beekeepers.

Nearly a decade later, Amos Ives Root, in Medina, Ohio, saw the benefits of this new design and, though a jewelry manufacturer at the time, began making beekeeping equipment in his factory. He quickly expanded his operation and became the largest producer of beekeeping equipment in the world.

Manufacturing was hitting its stride about then, when innovations such as electricity, rail transport, and rapid communication merged, allowing manufacturers to advertise their products to a wide audience, cheaply mass-produce the necessary items, and then reliably transport them to customers. The heyday of modern beekeeping had arrived.

Though rail transport reduced costs and delivery time, sending fully assembled hives by rail remained inefficient. (The Root Company was sending, frankly, boxed Ohio air to many distant locations.) Thus, they stopped assembling hives at the manufacturing plant and began sending them unassembled. Many more hives could be sent in a single rail car to be assembled by the customer on arrival.

At this time the United States still had a rural economy and the high cost of purchasing assembled beekeeping equipment was formidable. This further encouraged other manufacturers (who started a few years after Root) to produce only knocked down equipment. The labor costs were shouldered by the end user, who spent time rather than money assembling all those pieces.

An experienced assembler who has all the necessary tools can put together a four-box hive, including frames, in about four hours. And, once assembled, it needs two coats of paint to protect it from the weather.

A first-timer, with most of the tools, could do the same thing in, maybe, two days. For someone with only a passing interest in woodworking and with minimal tools, the task could take a week. If you're not exactly sure where your hammer is—right now—you are probably one of these people.

There were, essentially, no options 100 years ago. If you wanted to keep bees, you had to spend the time putting all those pieces together. For some, this is the best part of having bees. In fact, some beekeepers revel in starting from scratch and making their own equipment. They own, it must be noted, workshops that rival the one you see on a certain public television woodworking show.

Don't get me wrong. There can be an untapped, self-fulfilling satisfaction in working with fresh-cut wood, fragrant beeswax sheets, and the pleasant hours spent alone, or with a partner, in the aromatic assembly of hives. You may discover this joy in the journey—and while those pieces wait to be assembled, it seems like time stands still.

But these days, the journey isn't the goal for many people. It's having bees in the garden. This is where technology, labor, and the eternal press of time come together. There's now a full range of assembly choices, ranging from the traditional build-it-yourself kits to painted, fully assembled hives. If you choose the traditional route and assemble your own beekeeping equipment, be forewarned that the assembly instructions that accompany these kits are often woefully inadequate. But then, so are the typical instructions for assembling a propane gas grill. Either way, this book will explain it all.

Today's manufacturers use plastic and wooden beehive parts and assemble everything at the factory.

But First, a Little History

The history of keeping bees is rich, varied, silly, dangerous, over-flowing with keen insights, and weighed down by greed and ignorance. Thus it is sadly and wonderfully little different than any other practice in any other time. Its history has been, however, incredibly well recorded.

Given the immense volume of documentation that exists, we explore here only those events that were extraordinary in their advancement. The ambiguities of our craft and the dates of their discoveries can be cataloged by scholars and historians at another time.

For eons people didn't keep bees, they simply took the bees' honey. They found honey bees in forests and felled the trees they inhabited. They found bees in caves and robbed them of their bounty. Wherever these ancient people found bees they took what they wanted . . . paying the heavy price of stings. They probably discovered by accident (in a nighttime raid) that a burning torch with billowing smoke made the job less dangerous and more rewarding for the robbers, but less rewarding for the bees.

Eventually, it is supposed, an errant swarm found the confines of an overturned basket to their liking and took up residence. Protected from the elements and about the right size to build a nest, the trespassers prospered . . . until discovered by the basket owner in an unpleasant surprise for both.

Baskets evolved into basketlike *skeps* made of woven straw, twisted sticks covered in mud, and sometimes dung to keep the rain out. Still, because the bees fastened their delicate beeswax combs to the top and sides of these makeshift homes, the honey harvest was always destructive, and there was never a good outcome for the bees.

It soon became clear that this short-term gain was killing the proverbial goose that laid the golden egg. Developing a better, nondestructive way to harvest honey while still keeping the bees became imperative.

This is where apicultural sleuths take over, arguing the names of the explorers and the sequence of events leading to the eventual discovery of the removable comb, in (kind of) the following order:

- Removable top bars for the combs, then . . .
- Entire frames sitting on and sticking to the bottom of the boxes that had removable covers, then . . .
- An entirely removable comb, no longer attached to the top or the sides, surrounded by a frame of wood set apart from the sides and bottom, completely suspended within a wooden box: easy to re-move, and so easy to replace. Both bees and beekeepers rejoiced. Eureka!

So what you easily use today is the result of thousands of years of discovery, accident, and incident. You can now remove the cover and inner cover, pry up a slightly propolized frame, lift, remove, examine and replace it, leaving all unharmed and undestroyed. A series of acts taken for granted today, but only accomplished by decades of stings and discomfort, discovery and insight.

Straw skeps were the precursor to the moveable-frame hive. To harvest skeps, the bees were killed and the beeswax comb and honey were removed and processed by crushing the comb—an inefficient and inhumane production model. In the spring, swarms from the surviving skeps were installed in the empty containers to begin the process again. The development of the movable-frame hive allowed beekeepers to remove individual combs, harvest the honey without destroying the comb or kill-ing the bees, and return the comb to the bees to use again, which made commercial beekeeping possible, and the activity profitable and humane.

A New Concept

After a century and a half of very little change there's been a revolution in how beekeeping equipment is produced. The manufacturing technology was more evolution than revolution since these techniques are used for many products. The revolution came in the way the beekeeping industry began to think.

Some pieces of beekeeping equipment are always assembled by the manufacturer—covers, bottoms, and a few others. Preassembled and painted beehives are relatively new and have made beekeeping not only more enjoyable but more practical for beginners and seasoned professionals alike.

Professionals save time and money when ready-made hives, already on pallets, arrive at the beeyard ready for bees. (Labor is expensive.) For beginners and sideliners, the simple realities of having neither the right tools nor a practical place to use them are only a couple of reasons why assembling their own equipment has become so difficult. The garage—filled with cars, bikes, lawn mowers, garden tools, and the other stuff of life—generally does not have a built-in workshop. If used as a hive-assembly area, especially over a period of time, something has to give. When the task is complete, the "stuff of life" needs to be put back . . . somewhere. Basements are just as inconvenient to use as workshops, and few urban or suburban dwellers, which most of us are, have a barn or shop building out back. Basically, dedicating a space large enough to build what you'll need, and having all the woodworking tools to accomplish the task has become problematic and distracts from what hobbyists really want to do in the first place—keep bees in the garden.

So once you've wisely decided to use preassembled equipment, you'll find there are still more choices to make. For instance, what are your physical limitations? The common brood box—called a deep because of its height—when full of honey and bees weighs nearly 100 pounds (45 kg). This may be all right for weightlifters and sturdy teenagers to lift, but smaller boxes, called mediums, weigh in at about 60 pounds (27 kg) and are a better alternative for the average-strength beekeeper. Using the traditional setup, a typical beehive has two deep boxes and three, maybe as many as five, of the medium boxes. That's a lot of pieces to put together and a lot of lifting when they are full.

Beekeeping Fact:

There are hundreds of pieces in a beehive. Each box consists of four sides, dozens of nails, and frame supports. Each box holds eight or ten frames, consisting of six pieces, more than a dozen nails of different sizes, a sheet of foundation (beeswax or plastic), and wire (optional) to hold the beeswax sheet in place. There are four to eight boxes in each hive.

Let's simplify this. There are boxes available that hold only eight frames, instead of the ten in the traditional boxes. Better yet, they are available only in the medium size. Best, they come assembled. One of these, when full, weighs in at only 30 pounds (14 kg) or less. Weightlifters need not apply.

Wait, I'm not done yet—there's more to this tradition. I've named it the Zucchini Complex. Here's how it works: For springs eternal, gardeners have looked at their large, empty, fertile backyard spaces and imagined them overflowing with the perfect season's harvest. They see great, green, growing mounds of peppers and tomatoes, cucumbers, melons, radishes and beans, okra and greens, summer squash and winter squash, and carrots and corn. And every year, they plan and plot, order seeds and sets, and more.

I grew up in west-central Wisconsin, not far from Minneapolis. Though our neighbors' heritages were mixed, the common ground among them was dairy farming to earn a living and gardening to feed the family. Because of the blended European backgrounds, rutabagas and Roma tomatoes were grown side by side. But zucchinis were everywhere. They were fast growing, pest free, and in the spring, while still in the seed pack, nearly invisible.

You know the "August story." Every innocent zucchini seed, planted with love and care in May became a volcano of great, green fruit in August. If gardeners went away for a weekend, they grew to baseball-bat size. Three days of rain yielded three bushels of zucchini, with three zucchini to the bushel. We couldn't give them away because everybody already had too many. Mysterious mountains of zucchini appeared overnight on the side roads just outside of town. All this sprang from an innocent handful of seeds planted in May. That's the Zucchini Complex. Unfortunately, this complex also applies to beehives.

If you use traditional equipment, good management, and have even average weather, you'll end up with around 100 pounds (45 kg) or so of that wonderful liquid gold—honey— that your bees produce from each one of your hives. One hundred pounds—per hive. To look at it another way, that's nearly two 5-gallon (19 L) pails.

But this is more than tradition. It is the absolute goal of beekeepers everywhere. Those beekeepers, that is, who are intent on enjoyable beekeeping, sustained growth, a fair amount of labor and lifting, and profitable honey production.

But that's not our goal, not yet anyway.

The solution, of course, is obvious, whether for zucchini or honey. If your goal is not to produce record-breaking crops but, rather to learn the ropes, enjoy the process, and not be overwhelmed, then the best way to begin is to start with one, or better, a couple of hives in size eight rather than ten, and manage them so that monstrous honey crops don't overwhelm you with work and storage problems.

Promoting the concept of having bees that don't require hours and hours of work and that produce the size and type of crop that we can manage is the goal of this book.

A long-time friend of mine who is an experienced beekeeper, teacher of the craft, and keen observer of the people who keep bees, once said that people start keeping bees because of the bees, but they quit because of the honey. I'm going to make sure that doesn't happen to you.

CHAPTER 1 →Starting Right←

Keeping bees is an adventure, an avocation, and an investment, much like preparing for a garden. Considering the amount of sun, shade, and water drainage your yard provides, you must plan where your garden will be and how to prepare the soil. You must make an educated decision about what you can grow and what kind of care your crops will need. You will also need to be aware of harvest dates, and to avoid letting a lot of work go to waste, you'll need a plan for how to preserve the bounty. And, finally, you need to plan what needs to be done to put the land to rest for the off-season. The same planning process applies to beekeeping.

Getting Started

Your first step is to order as many beekeeping catalogs as you can find. They're free, and they contain a wealth of information. There are also magazines dedicated to beekeeping, and a free copy can be had for the asking. (See Resources on page 195.) Look particularly at those companies that offer preassembled products.

Next, read this book. Its chapters explore the biology, equipment, management, and seasonally organized activities of bees and beekeeping. It is important to become familiar with the seasonal routine of beekeeping. It is remarkably similar to scheduling your garden, but the specifics differ and need attention to master them.

Providing Water

Providing fresh water for bees is mandatory. A summer colony needs at least a quart (liter) of water every day, and as much as a gallon (4 L) when it's very warm. Making sure that water is continuously available in your yard will make your bees' lives easier, and it helps ensure that they do not wander where they are not welcome in search of water.

Water is as necessary to your bees as it is to your pets and to you. Whatever watering technique you choose for your bees, the goal is to provide a continuous supply of fresh water. This means while you are on vacation for a couple of weeks, when you get busy and forget to check, and especially when it's really, really hot—bees always need water. It is not likely that they will die, as insects are very industrious, but worse, they will leave your yard to find water elsewhere. Suddenly, lots of bees may appear in a child's swimming pool next door or in your neighbor's birdbaths. Outdoor pet water dishes become favorite watering stations for bees on the hunt for water. Bees

A beehive should be visually screened from your neighbors, the street, and, perhaps, even your family. The site should have some shade, lots of room to work, and a low-maintenance landscape. Notice that the white hive seen here is highly visible. Stark white beehives with little or no screening will draw attention to themselves. Though there is ample room to work around these hives, they are very visible and invite trouble. Out of sight and out of mind is the goal. This site doesn't work.

need water in the hive to help keep the colony cool on warm days, to dilute honey before they feed it to their young, and to liquefy honey that has crystallized in the comb. To make water accessible to bees, try the following:

- Float pieces of cork or small pieces of wood in pails of fresh water for the bees to rest on while drinking.
- Install a small pool or water garden, or have birdbaths that fill automatically when the water runs low.
- Set outside faucets to drip slowly (great for urban beekeepers), or hook up automatic pet or livestock waterers.

Join the Club

Find a local beekeeping club so you can connect with other local beekeepers. (See Resources, page 195.) Local club members have many things in common: weather, forage for their bees, zoning restrictions, sales opportunities, equipment, bee food and bee sources, similar pest and disease issues, and more. You can draw on the experiences of beekeeping veterans and learn a lot from the decisions, mistakes, and oversight of others.

Reap advice on when honey plants bloom in your area. Ask members what the local sources of nectar and pollen are and when they bloom. This will help you prepare your bees for the honey flows.

Some members have been keeping bees for years and years, while others like you are still climbing the learning curve. Long-time beekeepers have experienced many of the ups and downs you still have to deal with, and can ease you through them if you ask. They have learned enough to survive and prosper and can offer lots of good advice.

Always consider the perspective of the beekeeper offering advice. A beekeeper with hundreds of colonies has a different approach to most situations than someone with the same years of experience with only two colonies. Efficiency, scale, time, and profit may determine how that first beekeeper approaches the craft, while a love of nature, a fondness for woodworking, and enough honey to keep the pantry stocked informs the perspective of the second. You can learn equally from both if you consider each in their own particular context.

Beware of routine masquerading as knowledge. Success with bees over time indicates skill, knowledge, and hard work. But the skills, knowledge, and hard work may be due in part to doing most things the same way over a long time. This may work for one individual but may not be at all practical for you. Keep all this in mind when asking for and using advice.

You do not need to be an expert or experienced beekeeper to be an effective officer in your local club. New voices, fresh outlooks, new skills, and added contacts are generally welcome. Robert's Rules, taking minutes, and creating web pages or e-newsletters are just as important to the club as the skills necessary to introduce queens.

Regional associations can be great resources for beekeepers, too. Attending a variety of meetings broadens your exposure and experience, exposes you to other techniques and advances in management and pest control, and shares ideas that will benefit your local club. Larger, better funded associations (such as at the state level) may have the resources and contacts to provide the latest legislative information, and to influence laws, regulations, and funding that affect beekeeping and beekeepers. All groups benefit from your support, both financial or time- and work-wise. Be sure to take advantage of as many meaningful resources as you have the time for.

Most bee clubs meet at a member's beeyard, and new beekeepers can watch and work with experienced instructors. A club may have a mentor program that enables a new beekeeper to work with someone who has been keeping bees successfully for years.

Keeping Bees in Your Neighborhood

You probably know of neighborhoods that don't welcome weedy lawns or loose dogs or cats. Some areas also have restrictions on beekeeping. You need to find out about the ordinances of your city or town, because local zoning may limit your ability to keep bees. There are seldom regulations that do not allow any beehives on a suburban lot, but there are often specific, restrictive guidelines for managing them. And some places do strictly forbid having bees. Find out everything you can before beginning.

It is also important that you investigate your neighbors' take on your new hobby. It may be completely legal to have bees on your property, but if your neighbors don't tolerate your interest, you'll have to make some compromises. People's reactions to bees and beekeeping can be unpredictable. A few will be enthusiastic, most won't care one way or the other, and a few may have an extremely negative opinion of insects that sting and swarm. It's that last group you need to work with. If you are determined to keep bees, a little knowledge will go a long way, and there are some things you can do to allay a reluctant neighbor's concerns.

Often, the cause of a negative reaction from a neighbor is because of someone in the family being allergic to bee stings. Without being confrontational, you should find out if that person is really allergic to bees. Often people lump all flying insects together and yellow jackets or wasps may be the problem, while honey bees are actually not. It is true that a small percentage of the population does have a life-threatening allergic reaction to an insect sting (just as some have serious allergic reactions to peanuts or shellfish, for example). Most, however, have a temporary, normal reaction. Bee-sting symptoms include slight swelling at the site of the sting and a day or two of itching and redness. This is the typical response to a honey bee sting and should be expected. However, this book is not a medical text. You, your family, and your cautious neighbors should be very certain about allergic reactions to honey bee stings before you introduce a hive. Do not be alarmed, but do be careful.

Positioning Your Hives

Once you have considered everyone else's comfort level, it is a good idea to consider the comfort and happiness of your bees. Every family pet, including bees, needs a place that's protected from the afternoon sun and sudden showers and provides access to ample fresh water. Bees should be given the same consideration. Place colonies where they'll have a little protection from the hot afternoon sun, but don't place them in full shade. You'll see that the more sun your hive is exposed to, the better it is able to handle some pests. A bit of shade is good for both the bees and for you. All day sun is alright, but a bit of light afternoon shade also affords comfort for the beekeeper when working on a hot summer day, but not too much for too long.

Still in the Backyard

If having bees is legal where you live, but extenuating circumstances prevent them from being kept in the backyard, there are alternatives that can work.

"Beeyards" can be on back porches, with the hives cleverly disguised as furniture; on front porches, painted the same colors as the house and porch; and in storage sheds.

If you have a small yard, live on a corner, have a lot of foot traffic, or live near a school, check your roof. You may have a flat garage roof accessible from an upstairs window in your house. Problem solved.

Alternatively, place your beehives in a garage (with at least one window). You may work the hive from the inside, and your bees can easily come and go.

Hive Stands

A hive sitting on damp ground will always be damp inside, creating an unhealthy environment for bees. To keep your hives dry on the inside, set them on an above-ground platform, called a hive stand.

Before you choose a hive stand, consider that the closer your hive is to the ground, the more you'll have to bend and lift, and the more time you'll spend stooped over or on your knees as you work. This is an uncomfortable way to work, and it makes a good argument for using a raised hive stand. A 2' to 3' (0.6 to 0.9 m)-high stand strong enough to support at least

Cinder blocks are inexpensive, durable, and large enough to support your hives. Set cinder blocks directly on the ground, then place stout 2" x 4" (5 x 10 cm) or 2" x 6" (5 x 15 cm) boards, as shown, between the blocks and the hive. By the end of the season, this durable hive stand may be holding several hundred pounds of hive and honey.

500 pounds (227 kg) is ideal. You can build a simple stand using cement blocks and stout lumber. Another option is to make a stand completely from heavy lumber or railroad ties.

Build your hive stands large enough to set equipment and gear on while you work with the hives. If your colonies are placed at the recommended 2' to 3' (0.6 to 0.9 m) above ground, you'll need a spot to rest tools and equipment on during inspection. If your hive stand is small, you will be forced to set the equipment on the ground. Then you will have to bend over and lift parts all the way to the top of the hive to replace them. You will be better off creating an additional stand or additional room on one stand on which to set equipment. There is an old saying that is absolutely true: All beekeepers have bad backs, or will have. It is worth the extra planning to avoid the pain. A common technique is to build a stand long enough to hold three colonies comfortably, with about 2' (60 cm) or so between them. Then, put only two colonies on the stand with an empty spot between them. This space is where hive parts go when examining a colony, and retrieving them does not require bending all the way to the ground.

> **Tip:**
> **Keeping Hives**
> **Above the Fray**
>
> Keeping your hives high and dry offers protection from skunks. These fragrant visitors are notorious for eating bees.

Making Space

While putting everything together in your backyard—installing the visibility screens and your hive stands all at the right distance from your property line, and perhaps next to a building—you want to be careful not to box yourself in. Plan to have enough elbow room to allow you to move around the circumference of your colonies. This is especially true for the back of your colonies, where you will spend most of your time when working with the bees.

Grass and/or weeds are landscape elements that need to be taken into consideration as well. Left to grow, weeds can block the hive entrance, reducing ventilation and increasing the work of forager bees flying in and out of the hive. It is a good idea to cover a generous area around your hive stand with patio pavers, bark mulch, or another kind of weed barrier. Gravel or larger stones will work if you place a layer of plastic on the ground before installation. Even a patch of old carpeting will keep the weeds down and keep your feet from getting muddy in the spring or after a few days of rain. And a grow-free area cuts down on the chance of grass clippings being blown into the front door of a colony.

Good Fences Make Good Neighbors

Being a good neighbor includes doing as much as you can to reduce honey bee–neighbor interactions. Even if you have perfect neighbors, cautious management is an important part of your beekeeping activities and management plans. Here are some important considerations.

- Bees establish flight patterns when leaving and returning to their hive. You can manipulate that pattern so that when the bees leave the hive, they will fly high into the air and away, and then return at a high altitude, dropping directly down to the hive. There are several techniques for developing this flight pattern that will also enhance your landscape. Siting a fence, tall annual or perennial plants, a hedge of evergreens, or a building near the hives will help direct bees up and away from the hive. That same screen will also visually screen your hives from outsiders.

- Neutral-colored hives are less visible than stark white ones. Choose paint colors such as gray, brown, or military green, or use natural-looking wood preservatives. Any paint or stain formulation is safe for bees if you apply it to the exterior of the hive and allow it to dry before installing bees.

- Keep your colonies as far from your property line as possible, within any zoning setback restrictions.

- Avoid overpopulation. You should not have more than a couple of colonies on a typical lot of less than an acre.

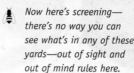

Now here's screening— there's no way you can see what's in any of these yards—out of sight and out of mind rules here.

Beeyards Other than Backyards

Sometimes keeping bees in the backyard and garden, the front lawn, or the roof is just not possible. There are a few essential rules to prioritize when deciding whether to set up your bees away from home.

First, your bees need to be nearby. If you spend most of your time getting there and getting home, then a trip back to retrieve that forgotten tool will too often end the day.

Also, it must be easy, safe, and legal to drive up beside your beeyard. If you have to carry your gear down a ditch, over a fence, or across a creek to your bees, getting stuck, ripping clothes, or getting wet will only happen a few times before it won't happen again.

Beekeepers too often put bees where they can rather than where they should. This is because, ultimately, we fail to do our homework. The location you choose needs to consider the well-being of you, your bees, and the surrounding community/environment.

Finding a good spot takes some work. Here are some considerations:

- You should be able to keep some tools at the site: smoker fuel, an old smoker, a few hive tools, a few supers, covers, bottoms, inner covers, and so on. Store the tools in a lidded container or a stack of bee boxes (with a lid) so they stay dry.
- There should be space for your vehicle to park close to where the bees will be, and room to turn around.
- Your beeyard should have safe, easy, anytime access, all year long, not just during the summer. Think "high and dry," especially in the spring.
- It is crucial that your bees be near a safe source of water, year round.
- Every beeyard should be surrounded by more than ample season-long sources of nectar and pollen.
- The beeyard should come complete with winter and summer wind breaks and great summer sun exposure, with the hives facing southeast, and it should not be in a cold-air drainage spot at the bottom of a hill.

The ground should be level, dry year-round, safe from cattle and other animals (who like to scratch on those hives), and out of sight from vandals. Your bees should be far enough from residences to be safe (from children, swimming pools, and so on.) but close enough so neighbors would probably notice anything going on that shouldn't be.

> **Tip:**
>
> Don't forget to use Google Earth and look around from that perspective. Even the free service, which generally shows somewhat dated photos, will show you more than you can imagine.

But if they are that close, all the good neighbor behaviors mentioned elsewhere should kick in.

Scout for locations by driving around your neighborhood, exploring the edge of town or country roads, even walking along roadsides. Ask friends and family, other beekeepers, farmers, ranchers, loggers, pilots, law enforcement agencies, and people in the recreation business for recommendations. Hunters, fishing enthusiasts, and bird-watchers often ask permission to access private property, too, and may have tips to share. And don't forget to scan the whole area with computer mapping programs that show fields, forests, roads, even gates and stone walls. Use both the street level function to look at the site from that perspective and the sky view to see what's over the hill and across the road.

If possible, take a whole season to evaluate a location before you decide to put your bees there. You may find that in summer your location may be high and dry and easily accessible, and in full sun in August. But come spring, a creek rises and you have to wait until May to get back there. Learn what the farming practices are of an area (crop rotations, pasture, plowing schedule, etc.). During the season look for available forage: which plants are blooming and in what quantity, and whether they are reliable. (Reminder: Your local beekeeping association contacts are invaluable for this information.)

Once you've identified three or four sites worth exploring, you need to get closer and take a good look. You need to find the owner and do an up close and personal inspection. Locating the owner may be a challenge; it could be a corporation without a face or a reasonable contact, a holding company for an estate, or a realty company looking to sell. Always contact the owner, and consider planning an in-person meeting.

Bring a map or printout from the computer program of the land you are looking at, along with a list of contacts for references to show that you are on the level and have a good reason to be seeking a location for your bees. The landowners need to know what to expect—your seasonal schedule, the kind of truck or car you have, how many colonies will be there, that a smoker will be going when you are there, and more than one person may be helping out. Be prepared to discuss insurance, security, accessibility, and more—and always volunteer to pay your rent in honey. This is why we too often put bees where we can, rather than where we want. You need a place for your bees that's good for you, good for your bees, and good for the landowner.

This beeyard meets most of the requirements of a good location. It is screened; shaded and in the sun; easily accessed and surrounded by room for driving; on level ground; away from seasonal water; and near ample nectar sources.

Extreme Urban Beekeeping

After three decades of decline during the struggle to cope with *Varroa* mites, the beekeeping industry has yet to climb out of its chemical fog and adopt industry-wide, reasonable integrated pest management (IPM) techniques. But bees that have some resistance or tolerance are becoming more available, and progress is beginning to be made.

Also during beekeeping's dark years, when there were fewer and fewer beekeepers anywhere, it was easy for urban areas—big and small cities and their suburban neighbors—to succumb to the pressures of those who were ignorant of the benefits of honey bees and regulate against them. Because there were fewer beekeepers, there were fewer voices to defend them. So during the decades *Varroa* mites were destroying beekeeping, they were ably assisted by misguided municipalities and many places became, simply, beekeeperless. There are no managed bees where there are no beekeepers. *Varroa* killed all of the bees, and governments killed the beekeepers.

But this has changed. Amid a growing awareness of habitat loss for all pollinators, coupled with the media-induced attention to honey bees brought on by the loss of pollinators of all kinds in epic proportions, the world woke up to the fact that the future of food was being threatened by the loss of honey bees and their keepers. The environmental and political action that resulted kindled a renewed empathy for all pollinators, including honey bees. Though we've always known it to be true, more people now see that it's good to be a beekeeper. Small farmers everywhere won a moral and productive victory when the rules in many cities and towns changed to once again allow bees to simply be.

Increased Regulations and Inspections

In most places, however, there are still restrictions on beekeeping. Permits that need renewal and cost money are usually part of the deal. Numbers are too, with a cap on the number of colonies allowed in a given space. Registration, training, permission from neighbors, description of housing, and other limitations often exist. Registration of hives with regulatory officials is usually required. But still, when you want to keep bees and you live in a city that now lets you, life is better for you, the bees, and the many plants that will benefit.

Over time, inspections and their requisite fees will probably become standard practice. Good neighbor beekeeping will take on a more official stance, because honey bees are being combined with many very close neighbors. The inspection programs are prepared to protect the city by making sure you are keeping your bees in a safe and secure manner. Hand in hand with inspections and inspectors is the permission for those inspectors to go onto the property the bees are on. Some locations require complete access to hives whenever the inspector wants it. Others work to make the inspection a teaching moment for the inspector and beekeeper. Most inspectors are fairly good natured and friendly. But the logistics of getting an inspector to hives in dense urban areas can be complicated. Just think: Where will that inspector park when he or she wants to get to the apiary on the rooftop of that popular hotel?

Extreme urban beekeeping requires extreme good neighbor beekeeping practices. The public, political, and legal landscapes have changed. The door has been opened, the welcome mat is out, and more and more places are willing to allow bees to be at home right next door. But you still have to pay attention to details, be on your (and your bees') best behavior, and remember that the rules for being an ambassador for good beekeeping have not been rescinded.

Urban beekeeping means that the bee–human interactions are more likely to occur than if you and your bees live in the country, suburbia, or small towns. Your goal is to minimize or eliminate these encounters. Remember, there may be more people afraid of bees than you can imagine, but even more think that what you are doing is fantastic.

Everything about keeping bees in the city or the suburbs is more focused, clear, dangerous, and exciting. The fundamental good neighbor practices mentioned earlier— good fences; water sources; neutral colored hives; using gentle bees; and temporary or permanent screens—still hold. But now the ramifications are greater if you aren't a good neighbor. Beekeeping in a large city is likely to be more challenging than a typical country location.

Urban beekeepers may have very small backyards . . .

. . . close to busy streets, and probably just off main thorough-fares so there's lots of foot traffic during some parts of the day . . .

. . . and just last fall this was all farm; there were no houses here at all. Really.

What Is Extreme Urban Beekeeping?

Extreme urban beekeeping encompasses many unique landscape and environmental factors. You may find: neighbors close enough to look in each other's windows, narrow lots, tall fences, garages tight to the houses, older homes, lots of good nectar-producing trees, small yards, tight landscapes, and gates. Also consider: front porches and tiny back decks, multistory apartment buildings, entire blocks of Brownstones connected all the way to the top, lots of heat-holding cement and asphalt, alleys (some with garbage cans—and garbage collectors, too), beehives on rooftops, pigeon coops, pigeons in the street, flower shops and green grocery stores on every block, bumper-to-bumper parking on the street, window boxes, fire escapes and front stoops with flower pots and barbeques, balconies with flowers and food, and kids. Your bees will find empty lots—some fenced in and some not—filled with the flotsam and jetsam of city life, including flowering weeds, garbage, old tires with water inside, boxes, and furniture. If your bees swarm, they can close businesses, streets, or entire city blocks until they are removed. (Who will remove them?) And you may have bees leave on their own to go who knows where.

But maybe you didn't move to a city and take up beekeeping. Urban beekeeping could also mean new development right next door—maybe you had a field of soybeans one day, but hundreds of people, golf courses, and swimming pools the next. It's different than dense city dwelling, but still a challenge.

Colony Growth

Beekeepers in densely populated urban settings routinely have fewer problems with pests and diseases because the number of colonies in one location is generally small and there are fewer beekeepers in the area, thus they are somewhat isolated from the problems of other bees. Although, with the increase of numbers of urban beekeepers this may be changing. Not surprisingly, losses from agricultural or homeowner pesticides are nearly zero (but not zero, as mosquito sprays and urban tree protection applications do exist). They lose fewer colonies to *Varroa*, nosema, and any of the common maladies that beset colonies in more traditional locations. Further, colonies become large enough to swarm and divide more often. There is a practical limit to the number of colonies you can have on one roof. Consider the logistics needed to handle the effects of colony growth (including all the honey you will harvest).

Swarms in the City

Swarm prevention becomes an even more important management tool in the city, for reasons other than having productive colonies. When a colony swarms in an urban area, it may land on the door of a business, a mailbox on a corner, even the side of a parked car, and the entire street may be closed to traffic and pedestrians. Most swarms will not garner much media attention unless their temporary resting stop is famous, dangerous, or photogenic. They may, however, attract unwanted attention from local authorities. Consider keeping your bees "out of sight, out of mind" in order to mitigate the potential negative attention. Remember, the fewer people that have access to your bees, the less likely they are to be vandalized or otherwise violated, too.

Up High on the Roof

"Out of sight, out of mind" is one of the reasons keeping bees on a rooftop is popular. People below never know there is a colony of honey bees up there.

Strong or constant wind can reduce bees' flight time and keep them at home. This requires you to specially plan the time you work your bees simply because it's easier working a colony when most of the inhabitants are away. Installing a windbreak against the prevailing wind or a two-sided screen will help the bees. And be careful that the hive stand legs do not poke through protective membranes on the roof.

Tip:
Colony Load and Bee Density

Bees are remarkably resilient to wind, sun, heat, cold, and other environmental stresses if they are provided with adequate housing, food, and protection. There is no reason that bees cannot survive on top of a ten-story, fifteen-story, or an even taller building. The amount of energy a bee needs to expend to fly that high when loaded with nectar and pollen, especially on a windy day, may be a stronger limiting factor. Pay special consideration to the wind with rooftop beekeeping. Shorter buildings that are in skyscraper canyons will experience more wind than rooftops in areas where all the buildings are about equal in height. If in doubt, sit on your rooftop for a while and experience the wind.

Controlling the number of colonies you have in any location (or bee density) in the city or country is another limiting factor. How many colonies can your area support? There's generally a wide variety of street trees planted in varietal clumps throughout the city. Most trees bloom in spring and early summer and are gone by midseason. What then? Parks and city plantings hold lots of flowers, and flower shops can supplement. (One flower shop with several varieties of sunflowers in cans on the sidewalk can feed more than 100 honey bees at a time.)

Take a look at online map services (such as Google Street View), walk the neighborhood, and examine vacant lots, which may have wildflowers blooming in late summer. Ultimately, an urban rooftop, balcony, or backyard may easily support two or three colonies, but ten may be a stretch. As anywhere, if the area is overpopulated with bees, the bees will not thrive. Another factor to consider is natural wild places within a city. Cities with rivers or lakes most often have a significant amount of undeveloped space unusable by people, but populated with flowering weeds all season long. Placing your bees close to these areas is definitely a plus.

Moving Your Equipment

Before installing hives on a rooftop, you must evaluate accessibility. You will need to get everything up to the roof, and then down. Before you order your equipment, measure all doors, windows, or other openings to be sure assembled equipment can pass through them. Even with adequate openings, getting a full-size colony off a roof can be a challenge. Outside ladders or fire escapes can be steep and narrow—which is less of an issue when moving empty, lightweight equipment but is potentially dangerous when removing equipment that is heavy, bulky, and full of bees. Also consider if your roof is accessible only through the apartment building's hallways, elevators, and lobbies. Removing a colony through communal spaces may pose problems such as errant bees, dripping honey, and cart maneuverability.

Bees on the roof need the same things as bees anywhere: continuous water. Bees on a hot roof need at least a half gallon (2 L), and as much as a gallon (4 L) a day in the summer.

Ground-Level Beekeeping

There are many other safe locations in a city where bees can be kept besides the roof—backyards, empty lots, alleys, decks, balconies, and porches. Any of these locations can attract attention if you don't take precautions. Common sense rules apply.

Watch flight patterns. When bees leave home, there's little incentive for them to fly higher than about 6' (1.8 m) unless there's a barrier in the way. If nothing is in the way they may run into people. Install a barrier or screen close enough that the bees are required to fly higher than 8' (2.4 m) almost immediately. This will minimize unwanted human contact with your bees.

Stay out of sight. Even though the city says "Yes, you can have bees in this city as long as you follow these rules," safety and common sense should rule the day. The population density of a city increases the likelihood that people may interfere, accidentally, mischievously, or maliciously.

Anywhere the colonies are should be out of sight. Neutral-colored hives work well, certainly better than white boxes, and living screens are effective for ground- or near-ground-level colonies. But remember, honey bee colonies do better in the sun. It keeps the bees warmer and drier, and makes it easier to dehydrate honey, plus neither *Varroa* mites nor small hive beetles do as well in a less humid environment. It's a trade-off. If the screens, fences, and gates are high enough

One of the greatest challenges to suburban or urban beekeeping is having bees near a neighbor's swimming pool. Fences may minimize contact, but the attraction of all that chlorinated water can be an irresistible force. Make sure your water source never dries up, and install screens to get the bees' flight path high above any swimmers.

to keep busy eyes away, they are probably high enough to keep sunlight off the bees most of the day. Try to locate your ground-resting colonies such that you only need two or three sides screened and some sun gets to the bees in the morning or afternoon. They'll be happier, and so will you.

Working Colonies

As with backyards, when working colonies in an urban setting you have to consider the people and pets that may be near you. Honey bees will defend their nest if they are threatened, and opening a colony is easily and often considered a threat. If your bees are close to where other people are, you want to open and work colonies when there are the fewest bees in the hive, which is midday during a honey flow when many of the older foragers are away, and many of the house bees are busy handling any nectar being brought back.

The section beginning on page 48 on working with the bees when examining a colony explains what you need to know no matter where your colonies are. Follow those guidelines and you and your bees will have a much better time together.

The opening of the hive can face any direction that's convenient for the traffic flow of people and bees. It's not critical which way it faces; just remember that your family uses your yard, and being able to keep your bees in check is important to everybody. Finding the best location for your hive will undoubtedly be a compromise between what you, your neighbors, your family, and your community consider important. Once you have decided on the best place for your hives, you have to consider the hive itself.

Bee Space

When honey bees move into a natural cavity, such as a hollow tree, they construct their nest by instinct, carefully producing the familiar beeswax combs that hang from the top of the cavity and attach to the sides for support, extending nearly to the floor of the cavity. To keep that spatial comfort zone called *bee space*, they leave just enough room between their combs so they can move from one comb to another, store honey, take care of their young, and have some place to rest when they aren't working or flying outside the hive. This space is not random. Measured, it is not less than ¼" (0.6 cm) and not more than ⅜" (1 cm). This distance does not vary between a natural cavity and a manmade beehive, and honey bees are unforgiving if presented with larger or smaller spaces. If there is a space in your hive larger than ⅜" (1 cm), the bees will fill it with beeswax comb in which to raise brood or store honey. If the space is smaller, they fill it with propolis. They do this to ensure there is no room in the nest for other creatures.

There are a couple of exceptions when it comes to comb building. Bees won't fill the space between the bottom board and the frames in the lowest box in a hive. They leave this space open to accommodate ventilation; the fresh air coming in the front door could not circulate through the hive if comb came all the way to the floor. Generally, honey bees also won't fill the space between the inner and outer cover. This rule is broken only when there is a lot of available food and not enough room in the hive to store it.

🐝 Bee space, shown here, is the space, or gap, between the top bars of frames in a hive. It is also the distance between the top of the frames and the top edge of the box. Bee space allows bees to walk about the hive. If the space is too large, the bees fill the space with honeycomb. If too small, they fill the space with propolis.

🐝 An obvious bee-space violation is pictured here. The bees had enough room to build comb and raise brood in the space between the top of this top bar and the bottom of the bottom bar above it.

Equipment: Tools of the Trade

Here are some important considerations to make while choosing your hives and personal gear.

Hives

We've already looked at the basics of the beehives you'll have. Seriously consider using preassembled, medium-depth, eight-frame boxes and appropriate frames. Amazingly, there are no standardized dimensions in the beekeeping industry. The dimensions of hives are not quite the same from one manufacturer to another. As a result, the parts of your hive may not quite fit together if you mix parts from different manufacturers.

If your boxes don't quite match, your bees will adjust. But their best efforts to hold the hive together in ill-fitting boxes work against your best efforts to take it all apart when checking on your bees. Sticky, runny, dripping honey from a broken burr comb (a free-form honeycomb built to bridge a gap between hive parts) makes a mess and will cause a great deal of excitement for your bees. Bees will weld ill-fitting boxes together (with a substance called propolis, which they make from plant resins) so that boxes become inseparable from adjacent boxes. The lesson: In the beginning, choose a supply company carefully and stick with it. Your first consideration should not be cost but ease and comfort for you and your bees.

To get a start in beekeeping, you'll need at least three eight-frame, medium-depth boxes for each colony. You'll soon need a couple more, but we'll explore those options later. Frames hang inside each box on a specially cut ledge, called a *rabbet*. Frames keep the combs organized inside your hive and allow you to easily and safely inspect your bees.

All boxes are similar, but there are minor design differences between manufacturers. The primary difference is how deep the rabbet is cut. Deep cuts allow frames to hang lower in the box than shallow cuts. When a box of frames is placed on top of another box of frames, there should be a sufficient bee space, 3/8" (1 cm) between the two boxes. If a frame hangs too low or too high when the boxes are combined, there will be too much or too little bee space between them. Either scenario makes manipulating the frames, the boxes, and your bees difficult. To avoid this situation, stick with a single supplier when adding or replacing equipment.

Pictured is an eight-frame hive, right out of the box. It has three medium supers, a telescoping cover, a screened bottom board, and a mouse guard in place.

An automatic watering device is an ideal way to provide water and not have to worry about the effects of a drought. (In winter, watering devices need to be unhooked and drained.)

Recommended: Preassembled Equipment

I always recommend that all starter equipment should be preassembled: boxes, tops, bottoms, frames, and everything you purchase. Assembling equipment is as inefficient now as it was 150 years ago. Only boxes and frames are routinely still sold unassembled—and the demand for even these continues to decline.

Much of the increased demand for assembled and painted equipment can be attributed to the commercial beekeeping industry. Mechanized, high-volume manufacturers are producing assembled and painted equipment more efficiently and less expensively than ever.

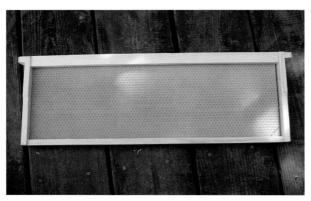

🐝 *This frame fits medium supers. It has a wooden exterior support that frames an embossed beeswax-covered sheet of plastic foundation in the center.*

🐝 *Make certain you have a screened bottom board. The one seen here has a removable tray beneath the screen to allow monitoring for Varroa mites. The tray can be either front or rear loading. Rear is best.*

Frames

Beehive frames are wooden or plastic rectangles that surround the comb. The outside provides support and maintains the rectangular shape of the frame. Bees build their honeycombs within the frame.

Brand-new frames start with the outside support and a sheet of what is called *foundation* within the frame. Foundation is a sheet that is embossed with the outline of the six-sided beeswax cells that bees build. One kind of foundation is made of pure beeswax, complete with the embossed cell outlines. These sheets are fragile and usually have vertical wires embedded in them for support. When assembling traditional frames with beeswax foundation, you frequently need to add horizontal wires for additional support. An alternative foundation is a sheet of plastic that is embossed like the beeswax sheets. These do not need supporting wires. There are also frames made completely of plastic. The outside support and the foundation inside are a single piece of molded plastic.

You can purchase unassembled wooden frames that come with beeswax or plastic foundation sheets. Assembled wooden frames are also available and come with plastic foundation. If the frames you purchase have plastic foundation, you will have to add beeswax coating (see page 34).

The suppliers who sell preassembled boxes also sell preassembled frames that fit in the boxes so that proper bee space is preserved. These are a good match and make setting up a hive much easier.

Bottom Boards

You'll need a floor for your hive. Although several styles are available, consider using ventilated bottom boards. Instead of having a solid wood bottom, these have a screen on the bottom, open to the space below the hive. Screened bottom boards are good for several reasons: The open bottom provides ample ventilation from top to bottom inside the hive, removing excess moist air and aiding the colony in temperature regulation, and an open floor allows the colony's debris to fall out rather than accumulate on the floor inside. You should, however, make sure there is some kind of a solid slide-in temporary floor.

Inner Covers

Set on top of the uppermost box is an inner cover. If the outer cover is the roof, the inner cover is the ceiling of your hive. It provides a buffer from the hot hive top in the summer and helps regulate air flow. There is an oblong hole in the center of the inner cover. Almost all inner covers are sold preassembled. They are often made from a sheet of masonite or a patterned paneling. These work but not well enough. They tend to sag as they age. However, some inner covers are made of several thin boards in a frame, which won't sag as they age. Find a source for the latter, as they are worth the search.

Additional items you'll need include a pail-type feeder, an entrance reducer, a bee brush, and a fume board. Each item is explained later in the book, according to when they are used during the season.

DIY Options

Increased efficiency and production to meet increased commercial demand has trickled down and reduced the price of assembled equipment for backyard beekeepers. This, of course, has increased demand and more manufacturers are figuring out how to efficiently produce, assemble, and paint beekeeping equipment.

A sizable minority of the beekeeping community sees assembling equipment as a rite of passage into the craft. And there's a skilled subset of beekeeping woodworkers who enjoy making their equipment from scratch and putting it together themselves.

Most manufacturers continue to provide inadequate assembly instructions, and sometimes no instructions at all. So, for those who wish to make the perfect wired frame, the perfectly square hive body, or the perfectly prewaxed plastic foundation, beginning on page 30 are the best assembly instructions you will find anywhere. Follow these guidelines and your equipment will last so long that your great grandchildren will be using it.

An inner cover sits on top of the uppermost super but beneath the outer cover. It has an oblong hole that allows ventilation, feeding, and escape. Most have a flat side and a recessed side, though some are identical on both sides. The notch on the narrow side provides an upper entrance when needed.

An assembled hive with the individual parts offset, showing from bottom to top: bottom board; three supers and frames; an inner cover; and a telescoping, metal-sheathed outer cover. An entrance reducer, which doubles as a mouse guard, rests on the outer cover.

Guide to Beehive Assembly

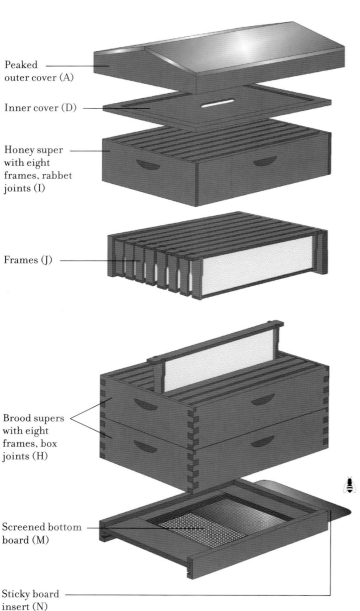

Peaked
outer cover (A)

Inner cover (D)

Honey super
with eight
frames, rabbet
joints (I)

Frames (J)

Brood supers
with eight
frames, box
joints (H)

Screened bottom
board (M)

Sticky board
insert (N)

Flat migratory
cover (B)

Telescoping outer
cover (C)

Feeding shim (F)

Hive top feeder
(E)

Escape board—
inverted (K)

Queen
excluder (G)

Pollen trap (L)

This illustration shows all of the parts and pieces of a modern beehive. There are three styles of covers: peaked, usually covered with a copper sheet (**A**), a flat migratory cover so colonies fit tightly together when being hauled on a truck (**B**), and the telescoping cover, which fits over the top of the hive (**C**). Colonies with migratory covers do not use inner covers (**D**), which sit directly beneath a telescoping cover and on top of the top super. Top feeders (**E**) allow bees to come up through the center slot and feed on sugar syrup you add to the trays on either side of the slot. A feeding shim (**F**) is for fondant, protein supplement patties, or as a spacer when treating your colony with formic acid pads. A queen excluder (**G**) is placed on top of the brood supers (**H**) and below the honey supers (**I**) to keep the queen from laying eggs in your honey supers. Frames (**J**) hang inside each super suspended by the extended ends on top of each frame. These ends are held in the groove on the inside of two ends of each super. Two joints are used to construct supers: rabbet joints and box joints. An escape board is shown upside down to illustrate the bees' one-way exit (**K**). A pollen trap (**L**) captures pollen that is used to feed to the bees later, or is sold by the beekeeper. At the very bottom is a screened bottom board (**M**) that has a sticky board insert (**N**) with the sticky board partially exposed.

Frame Assembly

For frame assembly, you may want a nailing jig and a form board, and you will need an embedding tool. Some use a brad driver for the small nails in the wedge.

Supply companies offer an assembly jig, which is designed to help you rapidly assemble ten frames at a time (page 33).

1. Twenty end bars, ten on each side, are held secure by spring-bound boards.

2. Glue (use a fast drying, reputable wood or all-purpose glue) is daubed on the joints at the bottom of the end bars; the bottom bars are then laid in and nailed.

3. The device is turned over, glue is daubed on the joints at the top of the end bars, and the top bars are laid in and nailed. Frames that are both glued and nailed or stapled last longer and do better in your hives. You can also use a staple gun loaded with long (usually 1½" [3.8 cm]) staples that have glue or resin on the sharp ends to secure them in place.

When lifting a heavy frame full of honey with your hive tool, you will exert several hundred foot pounds of pressure on the joint that holds the top bar to the end bar. Even more if the bottom of the frame is stuck to the top bar of the frame below. If the top bar is only nailed, it is likely to be pulled off. (See the diagram on page 32 for detailed nail placement.)

Wiring Frames

If you choose to use pure beeswax foundation sheets in your frames rather than plastic, they will need to be wired and secured to the frames you build.

Before wiring, purchase and insert metal eyelets that fit snugly in the predrilled holes in the end bars. These eyelets prevent the threaded wire from cutting into the soft pinewood of the frame. This keeps the wire from loosening, which would allow the sheet of beeswax foundation to sag.

When the eyelets are in place, begin threading the horizontal frame wires. Frame wire from suppliers is the right diameter and is made of stainless steel so it won't rust.

1. Thread the wire through the top hole on one side, run it across the frame and through the top hole on the other side, down the outside of the end bar and back through the next hole down. Run the wire back across and through the next hole down on the side you started on. (If you are doing this with deep frames, repeat this step.) You will end up with two (or four) horizontal wires running from side to side.

2. Pound a small nail halfway in on the edge of the end bar near where you started. Wrap the end of the wire around it to anchor it. If wiring a deep frame, pound another small nail halfway in the edge of the end bar near the hole where you finished. A single nail centered between holes works for smaller frames, as in the diagram. Pull the wire from the unfastened end as tight as you can (use pliers to assist if necessary) and secure it to the holding nail. Each wire should be tight enough to "ping" when plucked. When tight, pound in the nail(s) the rest of the way.

3. Remove the wedge from the bottom of the top bar with a pocket knife. Lay the foundation sheet behind the wires when lying on the form board. The bottom of the sheet fits in the groove in the bottom bar. Lay the sheet so the wire hooks face up when the foundation is laid down. Replace the wedge so it holds the foundation in by clamping on the hooks. Fasten the wedge to the top bar with tiny brads. Be sure to order the kind of foundation with both hooks and wire.

 A hybrid frame design is available both as knocked down or assembled. Foundation slips into the frame through the top slot and is held in place in the groove along the sides and in the bottom.

Guide to Frame Assembly

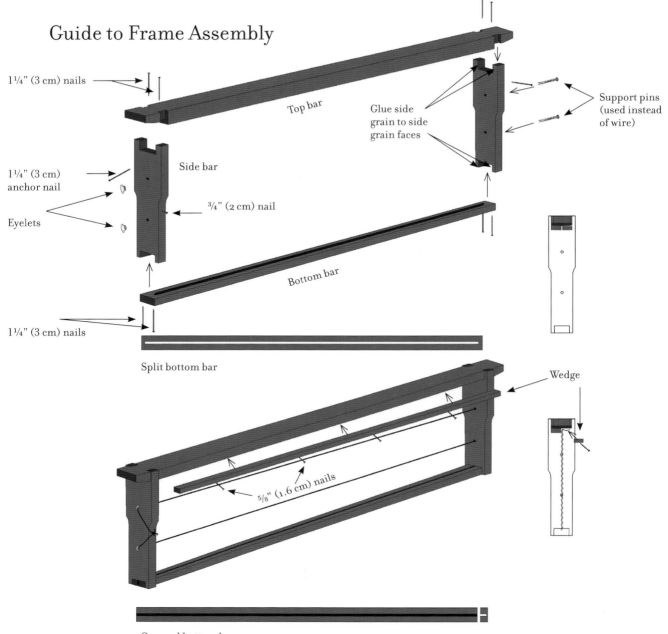

1¼" (3 cm) nails

Top bar

Glue side grain to side grain faces

Support pins (used instead of wire)

1¼" (3 cm) anchor nail

Side bar

¾" (2 cm) nail

Eyelets

Bottom bar

1¼" (3 cm) nails

Split bottom bar

Wedge

⅝" (1.6 cm) nails

Grooved bottom bar

🐝 *Frame assembly shows nails, nail sizes, eyelets, support pins, and other support structures (the wedge that holds the foundation to the top bar—a splitbottom bar which has a slot all the way through that the foundation fits through, and a grooved-bottom bar that the plastic foundation fits into).*

The wiring diagram shows where to fasten the wire on a nail to hold beeswax foundation in place.

The two most important nails in this illustration are the anchor nails that go through the end bar and into the top bar. These will hold the top bar and end bars together forever, in spite of the pressure you will put on this joint when lifting a frame from a sticky box full of honey.

Tip:
Plastic Frames

If you can choose the color of the plastic foundation, consider black. Many beginners have difficulty seeing very white eggs and young larvae in the bottom of cells on very white plastic foundation. The contrast between the black plastic and white eggs and larvae makes seeing them much easier.

Foundation should only be inserted into the frames two to three days before the frames are put in a colony because of its tendency to sag, even with good support from wires. Do the wiring early, but finish the foundation just before insertion.

When the frames are finished and the foundation is in, embed the wire into the wax. Use a form board, which is available from most suppliers. This board holds all sizes of frames so the sheet of foundation is supported from beneath. Push the wire down into the wax from the top side with an embedding tool. Without this kind of support the wire will push through the wax and ruin the sheet of foundation.

If you are using deep boxes for your brood, you do not need to wire those frames. Instead, use support pins (essentially a split rivet is inserted instead of wire, with each side of the split going on either side of the beeswax sheet).

If you are assembling your frames but using plastic foundation, you need to be certain you have the right frames to start with:

- Make sure the suppliers send you frames with grooved tops and bottoms for the plastic foundation. The grooves in traditional-style frames are slightly too small for plastic foundation.
- Look for models with a slot through the top bar and grooves on the sides and bottom to slide either wax or plastic foundation into.

Box Assembly

Box assembly is a straightforward task. Make sure the handles are all right side up and on the outside of the box before you glue and fasten them together. Two styles of joints are used to fasten the corners of the box together. Most common is the box or finger joint; less common is the lap, or rabbet joint. Both work well and are easy to assemble.

Joints can be glued or secured with screws so damaged sides can be replaced later. The box must be square when assembled. Some have metal frame rests in a few shapes to keep the bee space inside the box correct. If you don't install these rests, the space above the frames will be too big and the bees will fill them with comb, and the space below too small and propolis will always be in the way. Disregard the metal supports at your peril. Finish the assembled box with a primer coat and two coats of latex paint or a few coats of stain. The colors you choose can camouflage your colonies, or announce their presence.

Consider using a frame assembly jig that holds the pieces in place so you can hammer together ten frames at a time.

Three sizes of frames: (front) fully drawn shallow frame; (middle) medium wooden frame with plastic foundation; (back) all-plastic, one-piece deep frame and foundation with wax applied.

Beehives need protection from the elements. Two to three coats of paint or stain works well. Be sure to cover the exposed ends and the handholds, too.

Adding Beeswax to Plastic Foundation

You need to add more beeswax to plastic foundation so the bees have enough to work with when they first begin to make their comb.

The hardest part of this may be finding the two or three pounds (1 to 1.4 kg) of beeswax to cover forty medium frames, more for deep frames. It doesn't need to be candle grade but can be wax right out of a wax melter that hasn't been fine-filtered. To find wax, start with members of the local association, or check beekeeping equipment suppliers. This should be the only time you need to do this because you can collect your own wax in the future.

I have a two-burner hot plate. On this I place the bottom of an old roasting pan I got at a thrift store for pennies. I add enough water to fill it about a quarter to a third full, then I put in a couple of old cooking pans I got at the same store. I put enough wax in these to fill them about half full or so, place them in the water in the roasting pan and turn the heat on both burners. For photos of my setup, see "Melting Beeswax" on page 154. Your setup may be more sophisticated with perfect temperature control allowing you to be very precise and heat your wax to just the melting point of about 150°F (66°C). What you'll see, however, is that at about that temperature the wax is cool enough that it begins to harden on the surface around the edge of the pan it is being melted in. A yellow ring will form going all the way around. If the wax is too cool, a thin film of solid wax will begin to cover the whole surface. If it is too warm, there will be no ring at all. Wax that is too cool won't scrape off the brush, and if too hot it will simply fill the cavity at the bottom of the hexagons. Temperature is important, though not absolutely critical. Play with it a little to see what you need.

Dip a sponge brush into a pan of melted wax, and then paint the wax onto the plastic foundation. For more information on melting beeswax, see page 154.

Top Bar Hives

There has been a growth in keeping bees in boxes other than the traditional five-, eight-, or ten-frame boxes. Though there are several styles, we'll only focus on one because it typifies the advantages of these other boxes and highlights the differences in management practices required compared with Langstroth boxes.

First, European bees, those we commonly use, are cavity dwellers, and the cavity they choose has only a few requirements. There needs to be a ceiling to hang combs from and an entrance near the bottom of the combs. The cavity must be weatherproof, usually—but certainly not always—about 15 to 20 feet (4.5 to 6 m) off the ground, and about 2 cubic feet (0.06 cu m) or larger to accommodate a full-size nest at the height of the season. European bees make nests in the sides of houses, water meter boxes, small caves, upside down wheelbarrows, front porch columns, church steeples, and anywhere there is an accommodating space. Getting bees to live in an eight- or ten-frame hive is neither unique nor difficult. They are the right size, safe, and dry. Why hives that sit very near or on the ground rather than 20 feet (6 m) in the air are still attractive to bees remains a bit of a mystery, but size and comfort must overcome the lack of altitude because when we put European bees in our hives, they almost always stay there. So it should be no surprise that the shape of the box we put them in makes little difference as long as it meets the basics—volume, safety, and weatherproofing.

Top bar hives were developed as a compromise. Because of the climate and available forage, honey bees in Africa migrate seasonally. To take advantage of this behavior, beekeepers would prepare hollow logs 2 to 3 feet (0.6 to 1 m) long and about 1 to 1½ feet (0.3 to 0.5 m) in diameter to accommodate the bees when they arrived in their region. They closed one end of this hollow log with a removable door, and closed the other end but left an entrance hole; sometimes the whole end was left open. The logs were hung from trees on a rope used to raise and lower the log—you guessed it—15 to 20 feet (4.5 to 6 m) off the ground. Then the beekeeper simply waited.

Migrating swarms would arrive, find the logs, move in, build their combs, and fill them with brood, pollen, and honey. When the good season was over, the bees went in search of better weather. The beekeeper lowered the log, removed the door in back, and harvested the honey inside. It was a pretty straightforward process.

Enter beekeepers from other places with ideas to simplify and economize the process by making top bar hives. They reasoned that local beekeepers would need hives that were very easy and inexpensive to make because of the lack of available resources. So the hive must be made out of free local material that could be fashioned into basically the same shape

A typical top bar hive, this one with attached legs.

Another style top bar hive sitting on a small table, but without legs and easily transported.

as a log. That was easy: sticks and rope covered in mud. The same design could be used with hollow logs.

But instead putting a door on one end and a hole on the other, they made the man-made log such that they could raise the top to inspect and harvest what was inside, rather than come in from the rear. The goal was to enable the beekeepers to examine the hive and harvest the honey without destroying the comb and the bees in the process. So they removed the top third or so of this hollow log, lying on its side, and left a U-shaped cavity. To cover the opening, they placed slats across the log to give the bees a ready place to fasten their combs (the ceiling). These slats, now top bars, became the top of the hive, so they had to be made such to keep weather out. Then the beekeepers got clever and made the sides slant inward just a bit to accommodate the curve of naturally built comb. Otherwise the bees fastened their combs to the top and the walls inside the man-made log, just as they did with the original log. The thought was that the bees would hang their comb inside the log fastened to the bottom sides of the slats, or top bars, and wouldn't attach the comb to the sides or bottom of the man-made log. Individual combs could then be examined by lifting one of the slats, or top bars.

This design made management much easier and far less destructive. Putting top bars in boxes for the bees to build comb on wasn't a new technology. The main difference was that these newer models could be made of local, inexpensive materials and the shape modified to reduce comb being fastened to the inside wall of the cavity. And in fact for managing migrating bees in this environment, it worked pretty well. Beekeepers got easily constructed, moveable frames that could be transported to new locations.

The style and design of the hive evolved and was adapted to more moderate climates and much better behaved bees. However, because these hives are easy to make and all follow roughly the same pattern, as of yet there are no size standards. There also are very few manufacturers, and the additional pieces they make to fit their own equipment seldom fit other top bar hives. If you're handy you can make your own hive to fit a manufacturer's design and make additional or replacement parts to fit it. Or, simply design something that works for you from available material. So far, this is the most common experience.

Location is similar for any hive as far as screening, color, and flight paths. However, with top bar hives, being level takes on a whole new meaning. Strive to be nearly perfectly level side to side and end to end. Use a level to make sure. The bees will build level combs, and if the bars aren't level, the bees will build combs (called cross combs) attached to two or even three or four top bars, making removal essentially impossible or only with great destruction of comb. With hollow logs hanging in trees, the beekeeper didn't care because destruction at the end of the season was the plan. With these hives, saving at least the brood comb is the plan. Start level.

Top bar hives have a personality of their own, and there are several books available on building and managing them, plus there's a wondrous amount of information available on the Internet. As with boxes, get one already made and that you can get replacement parts for. Then follow the simple procedures outlined here to get started and keep going. Once you have a colony established, the guiding principles of basic biology and care are essentially the same, no matter the box.

Installing a Top Bar Hive

You install a package of bees in a top bar using techniques identical to those used in a Langstroth hive. First, define a space in the top bar cavity that is manageable for this many bees. Figure eight or ten top bars wide, with follower boards in place on either end and the entrance not quite in the middle of your space, or, if the entrance is on the end, place your follower board about that many frames away from the front door. (A follower board is simply a piece of wood or plastic supported by a top bar that fits snuggly to the sides and bottom of the hive, effectively keeping bees from passing into the space on the other side of the board. They are used to define or enclose smaller spaces inside the larger cavity of the top bar hive.) In a hive with top bars separated by bee space the food can be placed directly above the cluster, but top bars in a top bar hive seal the roof so a feeder can't be placed there, thus food must be under the top bars and within the cavity space with the bees. One way is to attach a slab of fondant to a follower board and snug it up close to the queen cage, no more than two or three frames away at first. If the weather turns cold, the food and the queen will be in the same place and bees won't have to abandon either. You can use a specially made frame, similar to a follower board but only a frame, that holds fondant or sugar candy feed.

Go through the ritual of package installation by removing the cover, the feeder can, and the queen cage. Attach the queen cage to one of the top bars so it is touching the bar near the feeder (you may need help attaching this, so bring additional string or rubber bands). Keep at least a couple of top bars between the queen and the follower board, but close enough that the cluster can touch both.

Remove half of the top bars from one end and dump the bees into the cavity, replace the top bars so the cavity is enclosed, and you're done. Check in a day or so to make sure the queen cage is still in its place, but be in and out in a couple of minutes. As with all queens, give her a week or so to become familiar with the bees, and if their behavior appears nonaggressive after that, remove the cork from the cage and let them release her.

During this approximately ten-day period, make sure the comb being built isn't cross comb, that is, attached to two adjacent top bars. A thin, long-bladed knife is a common, useful tool here to slide down the side to unfasten comb from the walls, but you'll need to use your hive tool to fix comb fastened to two bars. You may have to do this once, maybe even twice, but after that the bees usually leave it that way.

All new combs need to be inspected. I begin by removing a couple of top bars behind the follower board (the bees are on the other side of the follower board), above the empty space so I know there is no comb attached to it. Then I can slowly and carefully slide the follower board away from the adjacent top bar, first looking to see if there is comb attached to it from that bar. If there's no comb attached I move the follower board away, exposing the inside of the cavity being used by the bees. Use extreme caution when lifting a top bar by carefully separating it from the bar next to it. New comb is very fragile and if attached at all to the side of another comb, it will pull right off the top bar when you remove it and fall below. Try to always have an unused top bar next to the follower board when you begin your examinations so you can get in and look before you lift. This technique is essentially identical to examining a box of frames by starting on one end with no drawn comb on the foundation, removing the first one, then sliding each frame into the empty spot left by the removed frame.

When the queen is released and laying, comb building speeds up to accommodate brood rearing and pollen and honey storage. Check frequently for crooked combs, combs that fall, combs that stick to the sides, and cross combs that go off in random directions.

When a comb falls, and one or more will, you have to remove it. Simply reach in and lift it out, being sure to have a place to put it once it's out. If it's small, simply put it in the wax bucket you always have in the beeyard with you. But if it's large, with honey or brood, you'll find that a 5-gallon (19 L) pail lid works well because it is large and strong enough to accommodate almost any comb and is somewhat concave to hold any liquid honey. If reaching in doesn't work—it can be a bit unnerving—dedicate a kitchen spatula or long-handled spoon for the job. In fact, you will find that simply leaving these new-found beekeeping tools, along with that long-bladed knife and even a hive tool, right in the hive on the top bars saves you a lot of time all season long. Then scooping out the comb is easy. Don't worry about the honey or brood that falls out because the bees will make short work of those. However, you now have a broken piece of comb attached to the top bar. You can remove all of it, letting the bees rebuild completely, or let them rebuild from the break. The strength of the repaired comb at the break is about the same as unbroken comb, and letting the bees rebuild is okay.

New combs are not very sturdy for several weeks and, if laden with honey or brood and tipped the wrong way, will separate from the top bar. Never flip a comb the way you would if it had a frame around it because a new, honey- or brood-laden comb is not strong enough to support itself and will break off and be ruined. Rather, carefully turn it 90 degrees so the top bar itself is vertical rather than horizontal, then rotate the whole comb clockwise so you see the other side of the comb, and return the top bar to the horizontal position.

🐝 The parts of a two-pail system to extract honey from a top bar comb. The top pail has the bottom third removed and a metal mesh attached. The comb goes into the top pail. The nylon filter fits over the bottom of the top pail and is held in place with the ring cut from the portion of the pail that was removed. The bottom pail has a gate to empty the honey for bottling. This set up was designed by our friend Walt Dahlgen.

🐝 The setup with broken pieces of comb ready to mash. Use a vegetable masher. When done, let the strainer drain into the bottom pail, and remove and process the wax left in the top pail.

🐝 Lay combs on wire mesh at the bottom of the pail and mash with a potato masher. The wax stays above and the liquid honey stays below.

🐝 When finished, drain liquid honey out of the pail.

You can examine the opposite side with the comb above the top bar or twist to return the top bar above the comb. The point is to keep the weight of the comb resting on the top bar, or hanging vertically from the top bar, and to never hold the comb parallel to the ground. The wax attaching the comb to the bar is not strong enough to hold that much weight until it is much older. When replacing the comb, put it back the way you found it, reversing your motions as you go. One way to keep track is to simply mark one end of every top bar. That way the marked end always goes back where it came from.

Top Bar Harvesting

If the season is good, a colony will fill most of the cavity with comb attached to the top bars. The storage configuration depends on where the entrance is. When the entrance is on one end, the brood tends to be on the entrance end of the cavity, and honey is stored toward the rear of the cavity. If the entrance is toward the front the combs tend to have honey on the top quarter and brood in the bottom. The ratio of honey to brood increases the farther back you go, but then rapidly reverts to all honey comb. When the entrance is in the middle, however, brood tends to be in the center frames, honey on top and brood on the bottom. The ratio of honey to brood on a comb increases the farther you go from the center in both directions and rapidly ends up with all honey combs on both ends.

Harvesting, then, is fairly easy to do by simply removing honey-only laden combs. You can use them for cut comb honey or chunk honey, or you can squeeze the comb to release the honey and use the wax for candles or creams. You can make a strainer along the lines of the one designed by the creative Walt Dalghren using two 5-gallon (19 L) pails, one with the bottom third removed and mesh securely fastened in its place; this is then nested above the second pail. You can cut a ring from that bottom third and use it to suspend a nylon fabric filter. The combs are cut from the top bar and placed in the top pail. A vegetable masher is used to press the combs against the mesh, squeezing out the honey, which drips into the pail below. Let the wax drain, and either wash it to remove the remaining honey, or put it on top of another colony surrounded by an empty super and let the bees clean it for you. Be certain you aren't sharing disease when you do this. Melt the wax again the same way you would wax from a solar extractor, then filter and use. Because top bar harvesting is a relatively new practice, suppliers aren't yet producing harvesting equipment and you'll have to make your own. This shouldn't be the case for too long.

Winter in a Top Bar Hive

Wintering a top bar hive where winter is warm is fairly simple, just as it is with any hive. It is in the cold regions that wintering becomes a challenge. You can figure you'll need somewhere between 40 and 80 pounds (18 and 36 kg) of honey to get all those bees through the winter. A top bar comb plumb full of honey is, depending on size of the cavity, somewhere between 5 and 8 pounds (2.3 and 3.6 kg) of honey. And of course there are combs with the top half or so holding honey and the bottom half empty come fall when brood rearing slows. These are the combs you need to worry about. How much honey is in these half-full combs? That's the meat that will get your bees through the winter. So until you know how much is there, actually weigh some of these half-full combs and full combs. How much honey do your bees have?

If you don't have enough honey stored for winter, you'll need to feed your colony. You can make feeders to fit on front doors or make enough space inside the colony to accommodate a feeder jar in a holder. Fondant can be used fastened to the follower board or simply laid on the floor beneath the cluster. If there is a cover for the screened bottom board, be sure to replace it for the winter. Removing full combs of honey during the summer and storing for winter is a good idea if room allows. There's no better food. But in a top bar hive, bees cluster on the bottoms of the combs surrounding brood and eat the honey in the top part of the comb, moving up during the winter. Thus, all the honey in the world five combs away will not do much good. So, look for those combs with lots and lots of honey on top. Then, move combs full of honey close by so the bees can, when it warms, move to them and harvest the honey there, feeding the brood back in the broodnest, several combs away.

🐝 *A slab of fondant held on the follower board with rubber bands. When inserted in the hive, the fondant will be snugged right up next to the bees so they don't have to break cluster to get to the food.*

Protection is a must for these hives in cold areas. There are several ways to provide wind and cold protection, but interestingly, ventilation is not as important because the warmth of the interior of the hive keeps the bottoms of the top bars warm, so there is no condensation falling on the bees.

If your hive is off the ground on legs or a stand, lowering it to reduce wind exposure is a good idea. Placing it on the ground protected from moisture and cold with a sheet of poly insulation is a practical solution. Then surround it with bales of straw or wrap it in roofing paper for a wind block. If it has a peaked roof, place some kind of insulation in the space beneath the peak to keep the heat in and the cold out. The entrance, no matter where it is, serves as ventilation, although some models have a slim opening on the long side between the top bars and the roof that allows bees to escape that get trapped above the bars when you work the hive. If there isn't a cavity above the top bars, place insulation directly on the roof, and if necessary, protect that with roofing material. Hold it all in place with bungee cords or stout rope.

You may want to put a skirt around the legs and hive if they are one piece to keep the space under the legs enclosed and to protect the sides and bottom of the hive. Fasten the skirt material on the inside edges of the cavity so it stays fastened when you open the top, and completely wrap the hive, with the wrap secured both top and bottom so it can't be moved by wind or snow. If where you are has severe winters, you may want to fill the empty space below the hive with bales of straw to maintain a completely dead air space without any wind. Be sure to leave a small opening where the entrance is so the bees can get out for cleansing flights on warmer days. Entrances can be protected or reduced in size to protect them from direct winds. And don't forget screen or other entrance guards to protect them from mice.

One way to get food close to a cluster, whether for a new package or a winter cluster that is running out of food, is to place fondant vertically in the hive. You can simply place a slab of fondant on the follower board, holding it in place with rubber bands. You can make a feeder in the shape of the follower board as a frame and put fondant in the center held in place with wooden slats, rubber bands, or other means. Or you can simply make a follower board–shaped frame, empty in the middle, where the fondant can be put.

When the weather is warm enough, you can remove the insulation, return the hive to the stand, and examine the hive more fully to adjust brood, honey, and bees back to where you want them for the summer.

How to Examine a Frame in a Top Bar Hive

🐝 Move the follower board back and away from the next comb.

🐝 Keeping a few empty top bars between the follower board and the first full comb gives you working room when moving and raising frames.

🐝 After observing one side of a comb, raise one end of the top bar so the top bar is vertical and the comb remains as shown.

🐝 Turn the comb 180 degrees . . .

🐝 . . . and lower the top end of the top bar and examine the opposite side.

🐝 To return the comb to the same position, simply reverse your actions so the frame is in the same location.

🐝 If you do not separate comb attached to the side of the box before you remove the comb, you will have a mess. A break like this can spill honey, damage brood, or even cause the comb to separate from the top bar.

🐝 A typical comb. Brood is placed in the bottom two thirds or so of the comb, and honey is stored in the top third.

🐝 A top bar comb completely filled with honey. Because there is no standardization, you will have to weigh this comb to find out how much honey it holds.

🐝 When a honey comb breaks, honey will leak and collect on the bottom if it has nowhere to go. Bees will actually drown in this mess.

🐝 *One way to protect a top bar hive in winter is to wrap it in roofing paper and have it sit on foam insulation to keep it dry and warm. You can surround this with straw bales for additional wind protection.*

One task to consider doing during the winter is to make or obtain either another hive with the same dimensions so combs and top bars fit or to make a nuc, simply a shorter version of the original with the same dimensions. This so you can divide a healthy population and prevent swarming, while increasing your bee stock. Eventually you will have all same-size top bar hives similar to same-size boxes.

Top Bar Pest and Disease Control

Pest and disease control are needed no matter the box the bees are in, and this requires careful observation. Since the comb the bees suspend from the top bars is made to their standards rather than directed by premade foundation, some beekeepers claim that natural-size cells help the bees better combat pests and diseases. Not a lot of research has been done to support this, but there is some anecdotal evidence that it helps. My experience is that when using bees from the same supplier, they deal with pests and diseases about the same no matter the box. Your experience may differ, but in any case, be aware of the problems that can arise, and be prepared to help the bees when they do.

To control *Varroa*, a sticky board monitoring technique probably isn't going to work, even though some models have screened bottoms, so a sugar, ether, or alcohol roll will be required. Before using any chemical in any hive, please review the section beginning on page 108 on treating your hives for problems. Treating, if you choose to do so, with essential oil treatments will be difficult because there is no place above the bees to put the oils where the bees can get to them. To work, the bees have to remove the oil and, in the process, spread it in the hive. Plus, the heavier-than-air fumes have nowhere to go. Formic acid, too, will not work well because the seal between the top bars stops the fumes from reaching the bees below. However, suspending a formic strip from a top bar without comb has been tried with some success. Drone trapping is difficult because with comb built in a cavity like this, drone comb is on almost every brood frame rather than isolated on a particular drone comb frame. You could use one of the plastic foundation sheets here with drone cells already on it and cut to fit, but that challenges the idea of top bar hives using no foundation. It may be a solution, however. Miticide strips are probably the easiest to use, but they are certainly the most toxic choice. These are suspended from the bottom of a combless top bar in the center of the nest.

Top bar hives are considered more natural than other types of cavities we put bees in. If you do not put foundation in the frames of a Langstroth hive the bees will fill them with comb of their choosing rather than adhering, more or less, to the direction that foundation pushes them. Is that more natural?

Perhaps. Moreover, if a box-type hive isn't level, the bees will build cross comb in those also.

I see little difference in type of containers, really. If you choose to keep bees, you assume the liability of food, shelter, safety, protection from pests and diseases, and the same humane treatment offered any livestock you have, whether bees, chickens, cats, dogs, or cattle. That remains the most important aspect of keeping bees: your responsibility to them as the keeper. No matter the box, no matter the location, no matter the goals.

Personal Gear

Now that you've outfitted the bees, it's time to outfit yourself.

Bee Suits

A bee suit is your uniform, your work clothes, what keeps you and your bees at a comfortable distance, and what keeps your clothes clean. To meet the needs of the individual beekeeper, the sophistication and variety of bee suits is first rate. You'll find that white is the most common color, but any light-colored suit is acceptable. Full suits cover you from head to foot but are quite warm in summer weather. An alternative is a bee jacket. These are cooler, but they don't keep your pants clean. The important thing to keep in mind when working with honey bees is that they are very protective of their home. When anything resembling a natural enemy approaches, such as a skunk, bear, or raccoon, they will feel threatened. These enemies have one thing in common—they are dark and fuzzy—so, wearing dark and fuzzy clothes near the hive is not a good idea. Whichever bee suit style you pick, keep it simple to start, and get one with a zipper-attached hood and veil. These offer good visibility, durability, and no opportunity for an errant bee to get inside. And because the veil is removable, you can try other head gear later without having to invest in a whole new suit.

When you're examining your colony, bees will land on your suit and your veil, and they'll walk on your hands. This isn't threatening behavior, but initially it can be distracting and a little disconcerting. Wearing gloves can remove that distraction. Most people wear gloves when they start keeping bees, and most quit wearing them after a while. The cardinal rule is to wear what makes you comfortable.

Gloves

You can buy heavy, stiff leather gloves, which are made for commercial beekeepers, but our goal—as hobbyists—is finesse, not hard labor. I recommend buying the thinnest, snuggest, most supple gloves you can find. A common style is made of thin, plastic-coated canvas material. Long cuffs, called *gauntlets*, are attached. Gauntlets slip over your long sleeves to keep bees from climbing into your sleeves.

Bee suits come in two styles: jackets and full coverage. Full-coverage suits protect your clothes from wax, honey, and propolis, and they also keep the bees out of places where you don't want them. Full suits are good for heavy-duty work. Jackets provide less protection than full-coverage suits. Plastic-covered gloves are commonly used, fairly durable, and moderately good for fine motor skills.

Cleaning Protective Gear

After you have worn your bee suit and your gloves for several colony examinations, the amount of venom and alarm pheromone begins to build up in the material. Frequent washing will eliminate these chemicals and reduce visits from guards when you work a colony. Wash these clothes in a separate load so that alarm pheromones don't contaminate your other clothes.

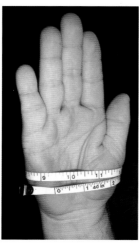

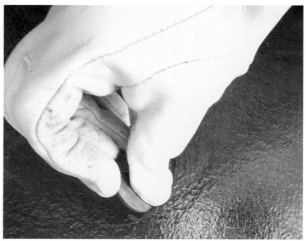

When buying sized gloves, measure the distance around the heel of your hand. The circumference in inches is the size of the glove you need. 5" (13 cm) is a size 5 glove.

All manner of gloves are available. From left to right: ranging from heavy duty leather, thin and pliable leather, canvas coated with plastic, cloth coated with plastic to thin dishwashing gloves.

Avoid squashing bees when manipulating hive parts during an inspection. Try picking up a quarter with a pair of heavy duty leather gloves. You have all the dexterity of a bulldozer. If you can't pick up a quarter with your gloves they have no place in a bee hive. But with thin or no gloves, you can feel a bee and not squash her. With heavy gloves you will squash bees, releasing alarm pheromone in the process.

Some suppliers sell gloves in exact sizes (not the traditional S, M, L, XL) and these, usually made of thin, soft leather, will fit best. This is especially important for the fingertips. Glove fingers that are too long make you clumsy and awkward, and it's difficult enough to be careful when moving frames. Regular rubber dishwashing gloves work well too, offering excellent dexterity when handling frames.

When you are ready to give them up, not wearing gloves is the best way to go for most beekeeping activities. But everyone has his or her own schedule for reaching that level of comfort. Eventually, you'll cut the worn-out fingers off an old pair of gloves to increase your dexterity. Then one day, you'll forget to put them on completely and not even notice.

Ankle Protection

Something not often thought about until it's too late is the gap between the tops of your shoes and the bottom of your pants. We seldom think of bees as being on the ground, but when you open a hive to lift out a frame or move boxes, bees will fall out. Most will fly away, but some won't. These are the bees you need to be aware of, because these bees will crawl—especially if the weather is cool or they are young and not used to flying. Sometimes, a lot of them will drop to the ground in

a bunch and crawl for a bit before they get their bearings and fly away. This is especially true if they land in grass or weeds, rather than on a smooth, flat surface. Bees that land on the ground naturally crawl up something. Usually their options are climbing up the hive stand or on your shoes. To avoid the latter, beekeeping suppliers sell elastic straps with hook-and-loop attachments that are easy to use.

Smokers and Fuel

A smoker is a beekeeper's best friend. A simple device that has changed little during the past 100 years, it is basically a metal can (called a *fire chamber*) with a hinged, removable, directional nozzle on the top; on the inside is located a grate near the bottom to keep ashes from blocking the air intake from the bellows, and the bellows. Only large- and small-size smokers are available, and the large model, no matter who makes it, is the better choice. Stainless steel models last longer than galvanized metal ones, but not much, and a protective shield on the outside of a smoker is there for a good reason. Buy a large, stainless steel model with a shield.

What do you burn in a smoker? Many fuels work well, but some are dangerous. Beekeeping supply companies offer fuel cylinders made of compressed cotton fibers and small pellets

Your smoker is indispensable when working with bees. Shown here is a good-size smoker. The beekeeper is wearing an attached-hood jacket and thin leather gloves for protection.

If you use any resinous smoker material, such as pine needles or shavings, creosote and ash and soot will build up inside the spout.

Periodically clean the inside of the spout on your smoker to remove buildup using your hive tool or other scraper.

And don't forget to clean out the opening so you get all the smoke you need.

Tip:
Elastic Straps

Long-legged bee suits with cuffs have elastic or closing straps that make crawling bees a nonissue. But, because a determined honey bee can make it an issue, having these elastic straps that can be wrapped around a wrist or ankle and secured with Velcro is a wise choice. Keep a pair in your back pocket.

---

of compressed sawdust. Have some of these available at all times, because there will be times when your other fuel is wet or you are out of your regular fuel source.

Many types of fuel are plentiful and free for the taking: Sawdust is one, chipped wood mulch is another, and pine needles are also wonderful. Dry, rotten wood—called *punk wood*—which is soft enough to crumble in your hands and can be collected during walks in the woods, is ideal. Small pieces of dry wood left over from a building project work well, too, as long as they fit into the fire chamber. Untreated jute burlap is good fuel, but be careful not to use synthetic burlap. Untreated twine from baled hay or straw also can be used, but beware—both burlap and twine are often treated with fungicides or other antirot chemicals so they don't disintegrate in wet weather. Make sure you are burning untreated materials. Don't use petroleum-based fire starters or gasoline. Bees are sensitive to chemicals, and the fumes from treated materials would kill your bees and probably cause flareups and other fire-safety problems in your smoker.

How Smoke Affects Bees

Thousands of years ago, someone figured out that if you have a large, burning torch with you when you go to rob a wild honey bee nest of its honey, you will have a much easier time of it. The smoke from the fire calms and quiets the bees somewhat while they are being robbed.

Several things happen when you puff a bit of smoke into your colony. Imagine the inside of your colony. It is pitch dark and about 95°F (35°C), with the humidity hovering around 90 percent. The primary form of communication in a honey bee colony is odor—when you puff in a bit of smoke, it masks odors and effectively shuts down communication, causing, understandably, quite a bit of confusion. The normal order is disrupted, and the chain of command is broken. This organizational breakdown allows a window of opportunity for a beekeeper to open the colony, examine what needs to be examined, do the work that needs to be done, and close up the hive before order is restored. When the smoke first enters the colony, some bees simply retreat from it. They run to the most distant part of the colony to escape the smoke. These are mostly house bees, which are too young to fly. Others head directly for the nearest stored honey and begin eating as fast as they can. Scientists suggest that this behavior occurs so that in the event that the colony needs to abandon its nest due to a fire, some bees will leave with a full load of supplies needed to sustain them while a new nest is constructed.

Some bees, however, seem not to be affected by smoke. These bees tend to be the guard bees that work on the periphery of the nest, and they are not as influenced by the communications that go on within the hive. Their tendency is to fly about, explore, and attend to the disruption.

However, even guard bees have some level of odor communication, which is disrupted by smoke. Ordinarily, when a honey bee senses danger in a hive, she emits an alarm pheromone—an odor. This pheromone has a banana odor that can even be detected by the human nose. When other bees detect the alarm pheromone, their instinct is to investigate the cause of the problem. If the threat is real, some bees may sting, which will release additional alarm pheromone into the mix. Bees will continue this activity until the threat is removed. With smoke present, however, even guard bees, which aren't in the colony and aren't affected by the disruption, will find it difficult to alarm other inhabitants. Although this situation may sound like a melee, it actually leaves the beekeeper to work in relative peace.

Smoking can be overdone, however, and once you have completely confused the colony, the disorienting effect is eventually negated, and, in fact, confused bees will begin flying no matter how much smoke you use.

Additional Equipment

There are some additional pieces of equipment that will make some things you do easier, and in most cases faster.

Burr Comb/Beeswax Collector

A beeswax collector can be as simple as a pail, can, or box to place beeswax scrapings into each time you examine your colony. Bees will put errant comb in places you don't want. Instead of discarding this valuable product, remove it and save it. Use your hive tool to slide under the comb and lift it up and out. Buy a small container that hangs on the side of your hive, make one, or simply use a small pail or can.

Cappings Scratcher

This very sharp, forklike tool is for removing cappings the uncapping knife cannot reach. A scratcher is also handy when looking for *Varroa* mites. (Open drone brood cells to inspect for *Varroa* mites.)

Maxant-Style Hive Tool

This common hive tool has a flat blade on one end and a rounded, 90-degree curved other end, which helps pry out frames from the sides or ends, wherever you can gain purchase. Slide the hooked end under the lug of the frame, use the base of the hook as the fulcrum on an adjacent frame or the side of the box, and easily lift the frame out. The flat end has a two-sided edge for separating supers or as a general scraping tool.

🐝 *A Maxant-style hive tool uses the notch on one side as a fulcrum to raise the hook, holding the frame lug on the other. A minimal amount of leverage lifts even the stickiest, heaviest frame.*

🐝 *A frame perch, or holder, keeps a frame close by, but out of the way when working in the colony. Various sizes are available. Be careful that the queen stays in the hive and doesn't ride along on the frame. She might fly away.*

Frame Perch

This device hangs on the side of a super you are working in. It gives you a place to hang a frame when you want to get it out of the box you are working in and keep it close, rather than set it next to the colony on the hive stand. Always remove a frame or two from the outside edge of the box you are working in—one with few bees, little honey, and no brood—before examining the rest of the frames. Carefully lift and examine each frame. If equally empty and beeless, place it on the perch or move it into the empty space created when you removed the first frame. This ensures that the frame with the queen stays in the box (especially if you didn't see her) so she doesn't fall off or fly away.

Queen Excluder

The queen excluder is a metal or plastic grid that covers the surface of an entire box. The grid is spaced such that workers can pass through the empty space but larger drones and the queen cannot. It is used to make sure the queen stays in the area of your colony used as a brood chamber and does not allow her to get into and lay eggs in the area of your colony used to store surplus honey.

When the bees allow brood to be produced in honey frames it darkens the wax in those frames, which darkens any honey stored there later. Plus, if there is brood in frames in the surplus honey storage supers, nurse bees won't leave the brood when you go to remove the honey with an escape board. Moreover, when examining the colony to judge the quality of the queen you'll know where to look, reducing the number of frames you will have to examine.

🐝 *A queen excluder in place on a colony: Nearly every model available violates bee space, so there is always a buildup of burr comb on it. Do not scrape the wires with your hive tool because you may bend the wires, enlarging the space and allowing the queen to sneak through. Place your excluder in a wax melter to remove the wax instead.*

Above is a variety of hive tools. Some are simply variations while others have a specific purpose. Choose a tool that fits your hand well, because you will spend most of the time you are working bees with a hive tool in your hand. (Note: The tool on the left with the padded area has a solid grip, but it is weighty. The second tool from the left is for cleaning between top bars. The third from the left is the Maxant-style hive tool.)

Two views of how to handle a frame and your hive tool. This keeps the tool always handy, but not in your way.

If you hold your hive tool like this, you can easily manipulate frames, even supers.

For reasons known only to the workers, sometimes they will not pass through the excluder, severely limiting the volume of the colony. The colony begins to think of itself as crowded and may swarm or they may stop collecting nectar because they perceive there is no place to store it. Sometimes it is said that queen excluders are instead honey excluders. This problem can be remedied by moving a comb or two with some honey from the box below to the box above the excluder, essentially telling the bees that it's okay to go up there.

Hive Tool

Beekeeping supply catalogs offer several styles of hive tools. The most utilitarian is the one that looks like a paint scraper with one end curved and the other end being broad, flat, and sharpened, and the standard 10" (25 cm) hive tool will provide the most leverage. Hive tools are inexpensive and, interestingly, easily lost. I recommend starting out with two. Other styles are designed for specialized tasks in a beehive, and you'll see their advantages when you've had some experience with the standard hive tool.

Tip:
Don't Be Fooled

You will see tools that look like a regular hive tool in most hardware stores, but there is a significant difference between these and the tools sold by beekeeping supply companies. These hardware-store paint scrapers are not tempered and will break when used to pry apart your supers or to remove frames. Hive tools are a hard-working part of your gear, so don't take chances with a tool not intended for this use.

🐝 *Inhive feeders fit in your brood boxes in place of a frame, and they are filled with sugar syrup. Choose a style with roughened sides or wire "ladders" so when bees go in to get syrup they can climb back out and not drown.*

🐝 *If collecting propolis (see page 80), a propolis trap is invaluable.*

🐝 *An escape board is an easy way to remove bees from a honey super. The bees exit through the wide end of the cone and are unable to find their way back through the narrow end. (The illustration on page 30 shows another type of escape board with a large hole leading out, and three tiny escape holes below into the super or brood chamber below.)*

Checklist of Equipment

Equipment for each hive should include:

- Screened bottom board with slot for sticky board for counting mites
- At least three assembled, medium-depth brood chambers, complete with assembled frames containing black plastic foundation
- At least two additional medium-depth supers for honey, which may be assembled regular honey supers, complete with assembled frames, foundation color optional
- Queen excluder (optional)
- Mouse guards and entrance reducers for the front door of the hive
- Inner cover
- Cover (Both styles, peaked or flat, work well; peaked are decorative but heavy and don't work as a platform for placing supers on when examining a colony.)
- Bee suit (with attached veil), gloves that fit well, ankle straps
- Hive tools—at least two
- Smoker and smoker fuel
- Hive stands to hold heavy hives
- Hive-top sugar-syrup feeder pail, jar, or hive-top feeder
- Books, magazines, and other beekeeping information
- Honey bees and a queen

The Bees

Now that you have your equipment, it's time to choose your bees.

Packages and Nucs, Colonies and Swarms

Beekeeping equipment is famously specialized. It is certain that your local hardware store, farm store, or discount home center won't have anything you need. This is especially true when shopping for the actual honey bees. Here's where being part of a local beekeeping club or having taken a beginner's beekeeping class is to your advantage. Knowing somebody who knows somebody who has whatever it is you'll need, whenever you need it, is the key to successful beekeeping.

There are several ways to obtain bees for your hives—two easy ways and a couple of other ways that are far more exciting and immediate. You can buy what is called a package of bees, which is simply a screened box containing honey bees, a queen, and a can of sugar syrup for food during travel, which will be shipped from a beekeeper who grows bees especially for this purpose. You will transfer these bees into your own hives, get them started, and keep them going as a colony. Or you can buy a small starter colony, referred to as a *nuc* (short for *nucleus*) colony, that you install in your own hive.

Another option is to purchase a full-size, ready-to-go colony of bees from another beekeeper. The advantage to this is that you minimize the risks of starting a small, somewhat vulnerable package or even a nuc, but the potential disadvantage is that you start out at full speed, without the breaking-in period that most beginning beekeepers need to establish their own comfort level with the craft.

Catching a swarm of bees is how some beekeepers get their start. This entails finding, capturing, bringing home, and hiving a swarm of honey bees. This activity is as exciting as beekeeping gets. (See "How to Catch a Swarm" on page 149.)

It's All in the Preparation

So far, we've looked at the tools you'll need to get started, reviewed the pieces and parts of hives, and planned where your hives will be located when they are up and running. We've also looked at your work gear—the protective suits and gloves, smokers, and hive tools—and where your bees will come from and how large the starter colony should be.

The old motto of always being prepared goes without saying, but I'll say it anyway. Start your preparations early; make sure you have everything you need; make all the helpful contacts you can; read the beekeeping catalogs, journals, and books, especially this book; and if at all possible, find a local club and take a starter course in beekeeping. And make certain that your neighbors and your family support your beekeeping aspirations. Now the adventure begins.

Buying Packaged Bees

Early spring arrives two or three months earlier in warmer regions than in more moderate and cooler regions, no matter where you live on the globe. People who live in warm regions and produce bees to sell start raising bees very early in the year, so they have them ready to sell when spring arrives later in cooler areas.

In order to do this, they remove some bees from their colonies every three weeks. They open a colony, find and remove the queen, and shake excess bees into a package (a screened cage) made especially for shipping live bees. The most commonly sold amount is a 3-pound (1.5 kg) package of bees, but 2- and 4-pound (1 and 2 kg) packages are also available. A 3-pound (1.5 kg) package is about the right amount for one eight- or ten-frame hive. There are about 3,500 live bees to a pound (455 g), so your 3-pound (1.5 kg) package will contain about 10,000 bees.

A can of sugar syrup supplies the bees with food for several days. A queen, snug in her own protective cage, is kept separate from the bees in the package because it takes a few days for the packaged bees to become acquainted with her. This complete package is shipped directly to a customer or a local supplier. Chapter 3 discusses how to get the bees and the queen from the package into your hive.

If you're lucky, somebody in your local club will have truckloads of packaged bees shipped directly to his or her place of business to sell in the spring. Check local suppliers before ordering, because it is best to buy locally. Find out what they are selling (the size of the package or nuc), the cost, the day the packages will be available (generally there is only a small window of opportunity—a weekend is common), and what choices for types of bees or queens you will have. Find out, too, where the suppliers are getting their bees and the queens and how long it will take for bees to be shipped. When it comes to price, the saying "you get what you pay for" is mostly true. If you live within a few hundred miles of primary suppliers, you may be able to buy directly, or have bees shipped to you through the mail. However, bees can be shipped only limited distances before the stresses of travel take their toll.

Buying Nucs, a Better Choice

There has been a change in thinking about starting with a nucleus colony instead of a package. Certainly, and unfortunately, the majority of beginning beekeepers get started with packages because for eons they were available in greater numbers than either nucs or full-size colonies.

And, generally, packages are less expensive than a nuc, but it is absolutely true that you get what you pay for. In short, if you can, get a nuc. If you can't, get a package.

A nuc (short for *nucleus*, meaning "small") is essentially a miniature, starter colony. Most nucs have five frames, but others may have three to six. They are produced in cardboard, plastic, or wood boxes that are not meant to be permanent. A nuc contains a laying queen, workers of all ages, open and sealed brood, drones of all ages and drone brood, stored honey and pollen, and all or most of the frames have drawn comb.

The nuc producer has taken much of the gamble out of starting a colony. When you purchase a nuc, its queen has been in that colony laying eggs for a minimum of a month and as long as several months. This amount of time lets the nuc producer evaluate her production and behavior, and replace her, on his time and dime, rather than you hoping the queen is a good queen, and the bees are healthy, and there's going to be enough food.

At the time of this writing there aren't nearly as many nuc producers as there are package producers, but their numbers are increasing every season. Your sources of nucs should have queens that have been producing for at least a month, or even better, several months. It is becoming increasingly popular to produce nucs by splitting large summer colonies in the previous season and furnishing each with a brand new queen going into winter. This produces a strong overwintered colony, a young vigorous queen, very few *Varroa* mites, and a great way to get into beekeeping. If possible, look for local nuc producers who produce their own local queens.

The one word of caution is equipment compatibility. Many, but not all, nuc producers use deep boxes rather than mediums for raising their nucs. You can get a deep box to accommodate a deep nuc and move them into mediums during the season.

How to Start with a Nuc Colony

You will need to transport your nuc home from the supplier. Generally it is secure and bees will not be leaking, but be prepared for a few stray bees in your car. If you are concerned, drive it home in a truck, or wear a veil.

Have your equipment ready before you leave. Review the information on package installation (page 94). You will need boxes ready on a hive stand, feeders and feed, entrance reducer, smoker and hive tool, and inner and outer cover.

No matter the weather, bring your nuc to your hive as soon as you are home. You cannot leave the bees confined in this

Commercial package producers produce enormous quantities of bees as well as their own queens. When a colony is large enough, they "shake" out several frames of bees into a funnel directly into a package, add a queen and feeder can, close the package, and ship it. Another way to gather the bees is to place a box containing bees and brood, and sometimes the queen, and smoke them down through the excluder into a screened box. When the screened box is nearly full after smoking or bouncing several boxes, the bees in the lower box are poured out into packages, weighed and are ready to ship. This technique is far, far more efficient than shaking a frame at a time.

small box. They will overheat without ventilation and perish. If it is a warm day, they will perish rapidly; if it is cool or cold, they will perish slowly. Do not hesitate or delay installing your nuc. If it is pouring rain, place your nuc in the exact spot the colony will eventually sit, open the front door so they can get some fresh air and fly, and leave them be until the weather improves. They will be fine and, when transferred, will feel perfectly at home. Put on your bee suit and veil, light your smoker, and remove the closures of your colony.

Remove the middle six frames in your box and set them alongside the box. Set the nuc alongside the colony. From one side of the colony puff a few puffs of smoke into the entrance, or raise the lid a tiny bit and puff smoke in the crack. Drop the lid, wait a minute, and repeat. Then, slowly remove the lid (and inner cover if it has one). Loosen the frame in the nuc that is closest to you with your hive tool. Slowly (with more smoke if necessary) lift it straight up, taking care not to bump it into the adjacent frame or side of the box. Keep the frame over the nuc. Slowly move it to the colony and place it in the box in the space you made next to the frame that is farthest from you. Repeat with the next frame, and then the next, and so on.

The frames in the box should be in the exact order they were in the nuc, but now are in the center of the hive. If you have a ten-frame box, replace five of the frames so there are nine in the box, and none of them are jammed tightly together. If you have eight, replace one on each side of the five you just put in.

Then, feed, feed, feed the nuc sugar syrup and protein supplement until the bees do not take it anymore. Let them adjust to their new home and new location for a few days before examining them. Afterward, the routine resembles that of a regular package examination (see page 97).

Buying a Full-Size Colony

Another way to get started with bees is to buy a full-size colony from another beekeeper. This approach makes you an instant beekeeper, but it also gives you all of the responsibilities that go along with being a beekeeper. You should consider a few things before taking this step. First, in the spring, full-size colonies will need to be managed for swarm control and monitored for pests and diseases, and will have a large population to deal with. There's no break-in period when you go down this road.

One other factor to consider when buying a full-size colony is that it belonged to someone else. Like buying anything used, you should have another, more experienced beekeeper or your local apiary inspector evaluate the colony for health and equipment quality before buying.

Types of Bees

All honey bees have a common ancestor, but their natural or man-assisted migrations have allowed for the development of species, or breeds, with adaptive traits. Honey bees now exist in all parts of the world except the two polar regions. Breeds have adapted to survive in deserts; during long, frigid winters; through tropical rainy and dry seasons; and in weather conditions between these extremes. The natural selection process has resulted in honey bees that are very skilled at living in cavities similar to traditional man-made hives, gathering and storing provisions to last during winter

Your nuc may be made of wood and have all the parts of a hive, only in size five instead of eight or ten.

when pollen and flower nectar is scarce or nonexistent, and choosing to swarm early in the food-rich spring, increasing their probability of establishing a new nest, storing food, and surviving future winters.

More than twenty subspecies of bees have been identified, and many of these have been tested by beekeepers for their ability to live in man-made hives, as well as their adaptability to the moderate climates of the world. Many subspecies have been abandoned by beekeepers because they possess undesirable traits, such as excessive swarming, poor food-storage traits, or extreme nest protection.

Italians (*Apis mellifera ligustica*)

Italians are by far the most common honey bee raised in the world. Having evolved on the moderate to semitropical Italian peninsula, Italian bees adapted to long summers and relatively mild winters. They begin their season's brood rearing in late winter and continue producing brood until

Bee Temperament

Gentle bees are easier and more fun to manage. Due to studied, deliberate breeding programs, the bees you buy today are gentler than the bees available twenty years ago. Every line of bees is different, though, and sometimes gentleness is more subdued. Experience is the best teacher when it comes to judging your bees' character, but here are some guidelines to look for when evaluating your bees for gentleness:

- Guard bees should not greet you before you get to the hive. They should stay in the hive or at the entrance. This doesn't include bees leaving to forage.
- In even a large colony you should not have many bees in the air after 10 minutes of having the colony open. A light puff of smoke should keep all the bees inside and between frames. There should be very few in the air when you remove the cover and inner cover.
- Bees should remain relatively still on top bars when you remove the inner cover.
- When a frame is lifted, the bees should remain calm and should not fly away, become agitated, nor run wildly on the comb, eventually falling off.
- Slow, easy movements should help you avoid any stings. Being stung should be the exception rather than the rule.
- Bees should not run or fly out of a super when it is removed from the hive and set aside.
- After you examine a colony, no bees should follow you more than a few steps from the hive. In addition, you have to work with your bees in such a way so they'll stay gentle. Follow the guidelines below to keep your actions to a minimum, and to engage your bees as little as possible.

- Only examine your colonies on sunny, wind-free, mild days (temperature between 65°F [18°C] and 98°F [37°C]) so as many bees as possible are out foraging, rather than staying at home, waiting for you.
- Absolutely avoid working colonies when it's cool, rainy, windy, cloudy, going to storm, or just finished storming.
- Don't start too early in the day, or too late in the evening. Between 10:00 a.m. and 4:00 p.m. is usually the best time because that's when the temperature is the warmest, the wind the least likely to be blowing, and the most bees out foraging.
- Always be gentle when opening the colony. Quick movements and loud, snapping sounds irritate the bees.
- Use enough smoke to make working the bees comfortable; but don't overdo it. Too much smoke will overdose the bees and soon they won't react to it at all.
- Keep your bee suit clean, and wash it often. The occasional sting on the suit will allow venom to build up, giving off an "alarming" odor to the bees.
- A manipulating cloth, which is a canvas and wire device that covers the tops of all of the frames of an open colony except the one you are working on, keeps bees contained and in the dark.

If your bees are not very gentle, and they begin to sting people and cause trouble, you can requeen the colony. (See page 60 for more information on requeening.)

---

the beginning of winter or later. Italians never really stop producing young, but they do slow down during the shortest days of the year. But not much.

Beekeepers living in southern climates have to deal with fewer management problems than their cousins in the north. There are nectar and pollen plants available during almost all of the bee's active months. But Italian bees kept in moderate and cool regions are challenged by a shorter growing season to make and store enough food to last through the long winters.

Package producers prefer Italian bees because they can start the rearing process early and raise lots of bees to sell. Beekeepers who pollinate crops for a living also like this trait because they can produce populous colonies in time to pollinate

early season crops. And Italians produce and store lots of honey when there is ample forage and good flying weather.

Italians are also attractive to beekeepers because they are not markedly protective of their hive. Italians are quiet on the comb when you remove and examine frames, they do not swarm excessively, and they do not produce great amounts of propolis.

Italians are yellow in color and have distinct dark brown or black stripes on their abdomens. The drones are mostly gold, with large golden abdomens lacking stripes. The queens are easily identified because they have a very large, orange-gold abdomen that is strikingly different from all the other bees in the colony.

Italian bees are generally yellow with brown or black stripes. Drones and queens have large, golden abdomens. The queen shown here also has a green spot on her thorax—not very common.

Caucasians are dark gray to black with lighter gray stripes on the abdomen. Queens and drones have dark gray to black abdomens.

Carniolan honey bees are dark with brownish to dark gray stripes. Queens and drones have nearly black abdomens. See the queen with the blue spot?

Carniolans (*Apis mellifera carnica*)

Carniolan honey bees developed in the northern part of southeastern Europe in the area of the Carniolan Alps, including parts of Austria, Slovenia, and areas north and east of that region. The mountainous terrain and somewhat unpredictable environment prepared these bees to survive cold winters and to react to quickly changing weather and seasons. As a result, they react quickly when favorable weather arrives in the spring, increasing their population rapidly and swarming early to take advantage of a short season. During the summer, they take advantage of the abundant food, but if drought or other unfavorable conditions arise they can slow their activity just as rapidly. When fall approaches, they slow their activity even more, and during the winter they survive with a small population and consume significantly less food than they do during the growing season.

Carniolans, unlike Italians, are dark in color. The workers are dark gray to black, with gray stripes on the abdomen. The queens are all black, and compared side by side, not as large as Italian queens. Drones are large and have all-black abdomens. These are the gentlest of all the honey bees. They are quiet on the comb when the beekeeper examines frames, and they tolerate typical beekeeper management duties. They also use propolis sparingly and tend to be a bit more forgiving in situations where burr comb would normally be used.

Caucasians (*Apis mellifera caucasica*)

Seldom used are Caucasian honey bees, developed in the Caucasus Mountains of Eastern Europe. They reproduce very slowly in the spring and react well to available resources during the summer. Like Carniolans, they respond to winter by reducing their population and using honey stores sparingly. But, because they build slowly, they swarm later in the spring than either their Italian or Carniolan cousins.

Caucasians are extremely gentle to work and are quiet on the comb when being examined. They also use propolis in every place you can imagine, which makes working your hive extremely difficult, but harvesting that particular hive commodity particularly attractive.

Caucasian workers are dark gray, with light gray stripes on their abdomens and sometimes brown spots. Queens and drones are dark, like Carniolans.

Other Bee Varieties

Varroa mites are parasitic on both adult and larval honey bees and are thought to have evolved in Asia, along with *Apis cerana*, the Asian honey bee. This coevolution over eons allowed *Apis cerana* to become resistant to, or tolerant of these mites.

The most common honey bees, *Apis mellifera*, originated in Africa and moved north to Europe and are today referred to as European honey bees. They regionalized into the other honey bee races (Italians, Carniolans, Caucasians, Macedonians, and others) and they did not coevolve with *Varroa* mites. When European bees moved (natural migration and man-assisted migration) to Asia, *Varroa* killed these invading European honey bees by the billions.

European bees continued to die until beekeepers began using pesticides lethal enough to kill the mites but not so lethal as to kill the bees. There are very few chemicals that fill this niche. European bees moved nearly everywhere around the world, and the mites (unintentionally) moved with them. At the time of this publication, *Varroa* mites can be found everywhere on the planet except Australia. The few chemicals that work are used extensively by beekeepers everywhere in the world. Even the best chemicals do not kill every mite in a treated hive. The survivors' offspring had the same resistance to these chemicals. Beekeepers and scientists tried other chemicals. They increased the dosages, the number of times during the year they treated, and tried various combinations of chemicals. The mites were unrelenting.

Not every colony of bees perished when attacked by mites. Most, probably 90-plus percent did, but those remaining would do fine without treatments. But did a colony survive because it was one of the very few that was resistant to mites or because it was treated? Honey bees were unable to develop resistance to the mites, as had happened with *Apis cerana*. So when the chemical *de jour* wore out, the bees continued to die, by the billions, until new chemicals came to be. The race continued.

There were two developments that broke this cycle. First, scientists working for the USDA theorized that the bees that had been exposed to *Varroa* mites the longest and had been exposed to no, or very few, *Varroa*-treating chemicals would have had the best chance to develop some level of resistance to these mites in the population. It turns out this place was in eastern Russia, where *Varroa* mites and European honey bees had been in contact the longest, and the beekeepers had the least access to the chemicals used in the rest of the world. And, when the scientists looked, sure enough, there were European honey bees living alongside *Varroa* mites and staying alive.

The bees were not a pure race. They were a mix of Carniolans, Italians, Caucasians, and even some Macedonian stock was present. There were several strains of this mixed stock. Some were well adapted to the late springs of the area so were very slow to build in the spring. Others were good honey producers but would build so rapidly that swarming was an issue each spring. Others were defensive, and some were as gentle as kittens.

They were unique in their heritage and they were unique in that most of them showed good resistance or tolerance to *Varroa* mites. Not perfect, but better than what had evolved in twenty-five years of exposure in the United States.

🐝 *This photo shows a female* Varroa *mite feeding on a pupating honey bee. She will raise one or two young on a worker honey bee, and two or three on a drone honey bee.*

Several of the best lines from several locations were developed under the direction of the USDA scientists and eventually many of these were released to the beekeeping industry. Soon after a group of dedicated breeders of these Russian lines of bees formed and developed a quality control and certification program. This gave a great deal of confidence and integrity to the program. Testing for purity is ongoing by USDA so buyers of Russian bees can be confident that they are buying certified Russians and not Russian hybrids. Moreover, selecting for desirable traits continues so the line is constantly improving.

Russian honey bees are not perfect, but they manage to stay alive with little or no chemical use in the hive. This was a quantum leap in *Varroa* control for beekeepers.

Russian Honey Bees

Beekeepers using Russian honey bees should have to treat for mites fewer times each season, and often no treatments at all are required.

They are very sensitive to the resources available in their environment. When food is plentiful they build their population rapidly and take advantage of the bounty. They are generally slower to build their population in the spring than Italians or Carniolans; they wait patiently for the first good nectar flow. This is a trait that many beekeepers find difficult to work with because they want their bees to build their populations early. Because of the rapid buildup that Russian bees exhibit later in the season, swarming can be an issue. Later in the season it is easier to control swarming because queens are more readily available to requeen the colony and the swarm, and the weather is generally more stable ensuring an easier time for both colonies.

Often, Russian hybrids, that is, pure Russian virgin queens, when open mated with local bees display an increased level

of defensiveness. This is not uncommon whenever mating pure strains with unknown strains. However, it has become increasingly obvious that mixing the hygienic behavior of Russian bees with locally adapted stock has produced bees that are more and more able to deal with *Varroa*.

Interestingly, tracheal mites are essentially nonexistent in Russian bees, and above-average hygienic behavior keeps colonies clean and diseases at bay.

Russian bees cease rearing brood earlier in the fall than most bees because of their environmental sensitivity. As a result they go into winter with fewer bees, and consume less honey during the winter than almost any line of bees.

Even though they are slower to build in the spring, they can rely more on their stored crop than having to collect the variable spring nectar or be fed.

Survivor Bees, For Lack of a Better Name

Some queen producers sell what they call *Survivor Queens* or *Resistant Queens*. They are hybrids of mixed heritage that are the result of deliberate breeding for *Varroa* mite resistance. Russian traits are common in these bees.

These bees are predictably productive, gentle, and extremely hygienic (in fact, this is proving to be the dominant trait in resistance and is being vigorously selected for). In fact, this hygienic behavior is fairly specific, aimed at finding and removing *Varroa* mites in the cells, under the cappings. Called Varroa Specific Hygiene, or VSH for short, it is one of the traits your queen producer should be actively selecting for. But your bees should have good wintering ability, be somewhat or very resistant to *Varroa* and other pests and diseases through activities other than hygienic behavior such as grooming, and most important, be repeatable, year after year. (This is similar to how the Russian program was developed; see page 53.)

In any event, survivor bees (by whatever name their breeders call them) have been successful in developing resistance to *Varroa*, and for the most part show the other

A typical Russian queen: Russians have a mixed heritage, so they may resemble Italians, be dark like Carniolans, or have a dusky, tiger tail abdomen like this one.

This is a survivor queen that has been marked by the beekeeper with green marker. It was selected from stock that has survived with some, but little, protection from Varroa mites. Be on the lookout for Varroa infestations, and be proactive in reducing any that show up by using the IPM techniques described on page 108.

traits beekeepers are looking for. One admirable trait that is generally common in these bees is that the queen is long-lived and productive for two, sometimes three years. Some breeders refer to this as the Survivor trait when compared with other bees with shorter-lived queens.

Unlike the Russian program, there is no certifying agency to confirm that indeed these bees are resistant to *Varroa* mites or any other malady. There are several university and USDA programs in the works that in time will make even this service available, but they are still in their infancy.

For now, when a queen producer says her bees are resistant to mites, you want to ask for data to support that, and perhaps someone to endorse those claims. A good producer will have both and not be reluctant to share them with you. It is worthwhile to find a good producer and try some of these. Though survivors are getting better and more available, without on-time IPM management, *Varroa* can and will damage some of the bees in your colonies. And damaged bees are not only stressed, but most likely to be virus carriers. So, even low *Varroa* counts need to be monitored.

Time Line for Beginning

The honey bee season follows the growing season, no matter where you live. In temperate areas, once the weather warms in spring and the days are long enough, plants grow and begin to bloom, and the honey bees will begin flying, foraging, and collecting nectar and pollen. A classic rule of thumb is to plan to have new bee packages arrive a week or so before the dandelions bloom where you live. If you don't know when

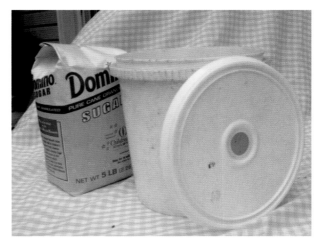

Sugar and a way to feed it to your bees is a critical part of getting your package or nuc off to a healthy start.

dandelions bloom, ask a local, experienced beekeeper (or gardener) when the bees ship, and start your plans with that date in mind. You'll need your equipment prepared before that date, and your hive stand and landscape screening set up.

In most years, ordering bees in late winter is good advice. The shortest day of the year is a good time to get an order in for the earlier packages and nucs, and you can order them as late as two months after that and still receive packages in the spring. After late spring, package shippers are reluctant to ship for fear the bees will overheat in transit.

CHAPTER 2 ⊹About Bees⊹

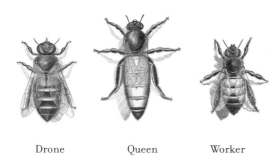

Drone   Queen   Worker

🐝 *Shown here are the drone (left); queen (center); and worker (right).*

Your honey bee colony follows a predictable cycle over the course of an entire season. To successfully manage it, you need a mental picture of what should be happening throughout the year, to help you time your visits, have the right equipment ready, and prevent problems.

You should also be familiar with the individuals within the colony. It is vital to understand the queen, the workers, and the drones, as well as how these individuals interact with each other, how they act and react as a group, and how they respond to their environment. Recognizing any situation that isn't normal is an important step in preventing problems or correcting them when they arise.

Let's start by looking at the individuals in the colony: the queen, the workers, and the drones. We'll explore their development and what each one does during the season. As we do this, we'll also examine the colony as a unit, as well as the bees' environment—including where they live, how seasonal changes affect them, and your interactions with them. In the next chapter, we'll bring all of this together and develop a predictable, seasonal plan that anticipates your activities and how to make beekeeping practical and enjoyable.

Worker Anatomy

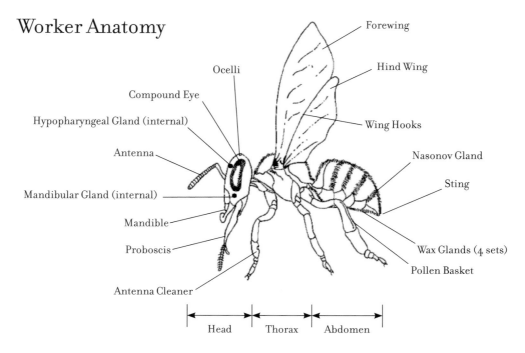

Ocelli
Compound Eye
Hypopharyngeal Gland (internal)
Antenna
Mandibular Gland (internal)
Mandible
Proboscis
Antenna Cleaner

Forewing
Hind Wing
Wing Hooks
Nasonov Gland
Sting
Wax Glands (4 sets)
Pollen Basket

Head | Thorax | Abdomen

🐝 *This illustration shows the basic body parts of a honey bee.*

The Queen

All bees begin as eggs laid by the queen of their colony. Eggs destined to become queens and those destined to become workers are identical at the egg stage.

A queen bee can fertilize eggs with sperm she received during mating that is stored in her spermatheca for her entire life, as she does not mate again. As the developing egg passes through her system, sperm is released and the egg is fertilized just before she places it in the cell. For three days, the egg develops within its shell. On the third day, the eggshell, or chorion, of the egg dissolves, releasing a tiny grublike larva.

Worker "house" bees immediately provide food for these tiny larvae, making a thousand or more visits each day to feed them. This food, for the first two and a half or three days is identical for both workers and queens. It's a rich, nutritious mix, called *royal jelly*, that the house bees produce from protein-rich pollen, carbohydrate-laden honey, and enzymes they produce in special glands. The house bees add the royal jelly to the cells, and one larva floats in each filled cell.

Larvae destined to be royalty see no change in this rich diet and continue to grow and develop. Workers-to-be, however, get a diet change on about day three. Their rations are downgraded in quantity and protein content, which keeps them from developing into queens. This difference, called *progressive provisioning*, allows the royal jelly–fed larvae to fully develop the reproductive organs and the hormone- and pheromone-producing glands necessary to fulfill their future

Queen honey bees have long, tapered abdomens and are larger than workers. They vary in color, depending on their race.

role as queens. They also mature faster than other bees, completing the egg-to-larva-to-pupa-to-adult cycle in only sixteen days, compared with the twenty-one days required for workers and twenty-four for drones. (See chart below.)

The cells in which queens are raised are different than worker cells. Because of the enriched diet, queen larvae are larger than worker larvae and require more room. Their cells either extend downward, filling the space between two adjacent combs, or hang below a frame. A queen cell is about the size and texture of a peanut shell with an opening at the bottom, making them easy to identify. The smaller worker larvae fit into the horizontal cells of the broodnest.

	EGG	LARVA	PUPA	TOTAL DEVELOPMENT TIME
Queen	3 days	5 ½ days	7 ½ days	16 days
Worker	3 days	6 days	12 days	21 days
Drone	3 days	6 ½ days	14 ½ days	24 days

Queen	
Worker	
Drone	

The diagram illustrates development time in days for queens, workers, and drones. Note that fluctuations in ambient temperature, colony population, nutritional needs, and other environmental events can alter these time frames. These are good estimates and should be used for planning colony management, but they are never exact.

🐝 *Because of their large size, queen cells are attached to the bottoms of frames, or fit between frames, and are easily visible.*

Queens are produced for a variety of reasons: to replace a queen lost through injury, in preparation for swarming, or to replace a failing, but still present, queen.

When a new queen is needed to replace an injured or failing queen, colonies almost never produce only one queen cell; they make as many as they can (if resources are limited) or as many as they want (if resources are ample). Queen cells can range in number from two or three to twenty or more and can be found on both sides of several frames. The process of producing multiple queen cells occurs over two to three days. Therefore, not all of the queen larvae are of the same age. The first queen to emerge destroys as many of the still-developing queens as she can find, eliminating the competition. She does this by chewing through the side of the queen cell and stinging the developing queen pupa inside. Sometimes, two or three queens emerge simultaneously, and they eventually meet and fight to the death, often with help from the workers. (Some of these workers and this new queen had the same mother—the queen that is being replaced—and the same father. These sister bees are more like each other than they are like their half sisters—those with the same mother but different father. They will assist their sister in this royal match and even help defeat a half-sister queen so their sister will rule.)

A colony generally tolerates only a single monarch, but on occasion, an older, failing queen and the triumphant daughter can coexist for a time. Sometimes, sister queens who emerge at the same time coexist without fighting. The common thread here is the close relationship and similar chemical cues they produce. In both cases, this is positive for the colony because of the increased egg-laying potential. Eventually, the older queen dies, or is killed by the workers. Retirement plans for queens are not very good.

For two to three days, the victorious, virginal queen continues to mature, feeding herself or being fed by house bees. Orientation flights near the colony begin after a week or so. The young, unmated monarch needs to learn the landmarks near the hive so that she can find her way back after a mating flight. Once she is comfortable with navigation, weather permitting, she starts mating. Queens hardly ever mate with the drones from their own colony (inbreeding could cause genetic problems in offspring). Instead, they take flight, looking for drones from other colonies. Drones and queens gather in places away from their respective colonies, called *drone congregation areas* (DCAs), mating 30' to 300' (9.1 to 91.4 m) in the air above open fields or forest clearings. Interestingly, drone congregational areas are in the same places every year, as long as the landscape and topography remain basically undisturbed. What is it about the edge of a forest or the open ground over a shopping center that draws drones and queens year after year? Last year's drones are not around to show this year's drones. How do they know?

A virgin queen emits an alluring come-hither pheromone during this flight, inviting a whole slew of drones to follow. The fastest drone catches her from behind, inspects her with his legs and antennae, and, if he deems her to be a potential mate, inserts his reproductive apparatus. The act stuns and seems to paralyze the drone. His body flips backward, leaving his mating organs still inside the queen. He falls and dies. These organs, called the *mating sign*, are removed by another drone if one catches the queen during this flight. She may mate with two, three, or five drones on each flight over several days. The last drone leaves his mating sign behind, and the workers remove it when the queen returns to her hive. You may see this on a landing board sometime.

Depending on the number of drones available—and, of course, the weather—a queen may make several mating flights within a few days. She may mate with as many as twenty or more drones or as few as five or six. Generally, the more the better, because it increases the amount of sperm available and the genetic diversity of the bees this queen will produce during her life.

Occasionally the queen will not mate because of an extended period of bad weather. After five or six days, she will be past mating age, so the colony will raise more queens, if eggs or very young larvae are available. If not, the colony may go queenless. This situation requires the attention of a beekeeper or the colony will perish.

When the queen's spermatheca is full, her mating days are over, and she begins life as a queen. Prior to mating and during her mating flights, queens are not treated like queens in the hive. They don't begin producing the colony-uniting pheromones until after mating. They do, however, have some pheromone control before mating. These pheromones inhibit further queen-cell production and the development of ovaries in workers, even though the egg-producing organs in her own abdomen—the ovaries and ovarioles—aren't completely matured until her mating begins. The queen appears to grow even larger now as these internal organs expand, but in reality, her abdomen is stretching to accommodate them.

Queens produce several complex pheromones, or distinct odors or perfumes, which are perceived by workers. Many of these chemicals are produced in glands located in the queen's head near the mandibles. According to bee scientists, at least seventeen compounds are produced in this volatile mix, often simply referred to as queen substance. Several other pheromones are produced by the queen in other glands. As the worker bees feed and groom the queen, they pick up minute amounts of these chemicals. Then, as they go about their other duties, they spread the chemicals throughout the hive, passing along scent cues that inhibit certain behaviors and strengthen the frequency and intensity of others. The most important message relayed by these chemicals is that there is a queen present.

In an unmanaged colony, barring injury or disease, a typical queen will remain productive for several growing seasons. As she ages, her sperm supply is continuously reduced, and her ability to produce all of the necessary pheromones for colony unity diminishes. Usually, the correct proportion of the many pheromones becomes unbalanced and some of the inhibitory behaviors are no longer contained.

There comes a time when the workers in a colony are able to detect that the appropriate pheromone level is no longer sustained. This situation can happen for several reasons, but overcrowding and supersedure are the most common. These two events elicit very different and distinct behaviors in the colony.

Swarming

Crowding often occurs as the growing season takes off, the days get longer, there's abundant forage available, the weather is favorable, the population of adult bees is large because the queen has been laying steadily, and the brood population is rapidly expanding, taking up much of the available space and giving off a whole hive full of pheromones. The colony is crowded with adults, more are on the way, there's little room

When a swarm of bees leaves its colony, it fills the air all around the colony. The swarm then heads for a nearby branch or other object on which to settle before moving to its final home.

to expand, and the external environment invites exploration. Add to this a queen that's at least a year old, maybe older, that is beginning to show her age just a bit and is producing overall less pheromones in total, and the ratios of those she does produce is different than when she was younger. Lots of bees, aging queen—the amount of queen pheromone per bee in a colony like this is reduced, and the dilution effect triggers the bees to focus their efforts on beginning the swarming process. This means that about half the workers in the colony change from a brood-rearing, foraging mode to one of slowing down production, packing up, and preparing to move.

One result of this situation is that young workers, those able to produce wax, begin constructing the base of large queen cells along the bottoms of the frames in the broodnest, building them so that they hang down from the bottom of the frame. These bases are called *queen cell cups* and you will often notice them along the bottoms of frames. Finding them may indicate the early stages of swarming plans. At the same time, some of the previously foraging bees begin to look for a habitat that would make an acceptable new home.

Because room for expansion in the colony is limited, the queen slows egg-laying behavior, is fed less food, and, within three to four days, stops laying eggs completely. A queen who isn't laying eggs loses weight and slims down because her ovarioles shrink. Her newfound slimness allows her to fly—something she hasn't done since her mating flights.

Her changed behavior becomes more intense over the five to six days that the new queen larvae are developing, and those workers that have become aware of the changes are no longer foraging but staying in the hive and gorging on honey—gathering provisions for the move. The final piece falls into place when the first of the queen larvae reaches pupating age and her cell is capped by the workers. This is the final signal for the existing queen to move to a new residence.

If the weather cooperates, the scout bees, which have been searching for a new habitat, and other workers begin racing around in the broodnest area, stirring up the colony. Suddenly the bees—including workers, a few drones, and the reigning queen—leave, pouring out the front door by the thousands.

Staying to keep the home fires burning is the remaining population. They continue to work and live in the colony as if nothing has happened, foraging, ripening and storing nectar and pollen, and tending to the new queen. Meanwhile, the departing swarm fills the air around the colony. It slowly organizes and heads for a nearby location, such as a tree branch or fence post, usually 50 yards (45.7 m) or so from the colony's former home.

Scout bees, those that have investigated possible new homes, join the waiting swarm and perform directional dances on the surface of the swarm to persuade more scout bees to visit the prospective locations. When one site draws more visitors than others, the scouts return and begin the mobilizing activity again. The swarm rises and heads toward its new home. There, new comb is constructed, foragers begin immediately to bring nectar and pollen home, and soon the queen begins to lay eggs. A new colony is complete. Of course it is much, much more complicated than that, and books have been written about this incredibly complex event and the behaviors before, during, and after a swarm.

Meanwhile, in the original colony a new queen has emerged and mated and is laying eggs. The colony continues as before, but it has had a break in its egg-laying schedule of between three and four weeks. This can be an advantage when it comes to *Varroa* control since there is little brood for them to reproduce on and their numbers will fall until brood rearing begins, which will be an additional week or more until cells are ready to be capped.

Supersedure

Supersedure, or replacement of the existing queen, occurs not when the colony is in swarm mode, but because either an emergency has occurred or the old queen is failing. Failing queens are not all that uncommon. Poorly mated queens seldom last very long and are replaced. So are queens with Nosema disease, as are those who have been damaged by virus or other diseases. A healthy queen is not guaranteed in your colony, and you need to be aware of both her health and performance.

Recognizing a Queenless Colony

Colonies that have lost their queen display some definite behaviors that will cue you in to the situation. These behaviors, however, can occasionally be noted when the colony is queenright (having a healthy, laying queen). It's not always perfectly clear what's going on, but a close examination usually shows that several queenless behaviors are occurring simultaneously. As soon as an hour after a queen disappears, the lack of her pheromonal presence is pretty well understood by all or most of the bees in the colony. Within a day, sometimes more or less, the bees will undertake emergency supersedure behaviors.

With the queen's pheromone signals vanishing, the colony becomes stressed, leaderless, and without direction. Behaviors include an increase in fanning behavior, seemingly to better distribute what regulating chemicals are remaining in the hive. This fanning is noisy—literally. You will notice the

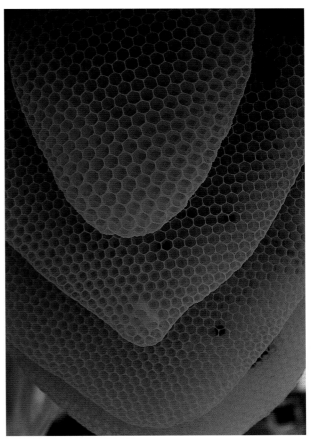

🐝 *The first thing bees do in their new colony is produce beeswax combs so that they have a place to store food and raise young.*

difference immediately when you remove the hive's cover. At the same time, you'll notice an increased defensive level from the guards at the front, and even from those who rise to meet you from the top. More bees in the air, louder sounds, and a generally agitated state typify a short-time queenless colony.

Other factors can also elicit these behaviors. A skunk or raccoon visit the previous evening can agitate the bees for most of the following day. A whiff of pesticide—not enough to kill lots of bees—can cause fanning, agitation, and defensiveness for several days. Opening the colony on a cool, rainy day or during an extreme dearth can cause defensiveness because there are more bees than usual at home. You'll have to explore a bit to be sure of the cause, but the sound of a short-time queenless colony is distinctive. If the colony has been queenless for a week or longer, other behaviors emerge, such as no eggs and supersedure cells— a sure sign of what's going on.

If the behavior continues for longer than a month, workers may start laying unfertilized eggs, foraging will be significantly reduced (when compared with other colonies in your yard), and there may be no drones. After a month, a colony can remain queenless if it simply wasn't able to raise a queen. Even if they did produce several supersedure queens, all can be lost while fighting for supremacy (it happens often enough to mention), or the winning queen can be eaten by a bird or become lost while out looking for a mate.

There are a lot of bumps in the road back to monarchy without your timely intervention.

Emergency Supersedure

One event that requires a colony to produce a new queen is the sudden death, loss, or severe injury of the current queen. Sudden death can occur if the queen is accidentally crushed by a beekeeper during a routine colony exam. Loss can also occur when a frame is removed, and startled by the sudden exposure to light, the queen flies off, looking for the warm and dark broodnest from which she was suddenly removed. Not a strong flyer, she can become lost, sometimes not far from home.

Any injury is likely to alter the queen's ability to lay eggs, produce pheromones, and eat. These deficiencies are immediately evident to the workers because of the constant attention the queen receives. They may or may not continue tending her, and in a time as short as one to three days, they react to the reduction or lack of egg-laying, queen substance, and her other pheromones. These events signal the beginning of queen-replacement behavior among the workers. Because of the urgency, this process is referred to as an emergency supersedure.

During swarm preparation, the colony receives a series of signals and reacts to each in turn, building up to the finale. Unlike during swarm preparation, workers don't make queen cell cups during an emergency supersedure, because no queen is available to lay eggs in them. Instead, the house

Supersedure queen cells are found on the face of this comb rather than the bottom because the bees have to choose an existing egg rather than prepare a special cell. The queen cell pictured here hangs between adjacent frames.

bees and those actively feeding the brood search for eggs or the youngest larvae they can find that are still feeding on royal jelly—the special diet fed to future queens, which allows complete development of their reproductive organs.

When eggs or royal jelly–fed larvae are located, the wax builders begin building queen-size cells for them. Because the egg or larvae could only be found in a regular, horizontal cell on a frame, the queen-size accommodations are built outside the frame's face and extend down and between adjacent frames. Several of these may be made by the colony if resources, such as food and larvae of the right age, are available.

Normal Supersedure

The second event that can trigger queen replacement is the normal aging of a queen. As the queen ages, the sperm she acquired when mating is slowly depleted and eventually is gone. As this occurs, she lays more and more drones (unfertilized eggs turn into drones) and fewer workers, creating an unbalanced hive population. Though a healthy, prosperous colony can afford to support a large population of drones, it is the workers that ensure colony advancement and survival.

To regain the balance, a fertile queen that produces mostly workers is needed. The colony will produce another queen in the same way as in an emergency supersedure, but with the insurance of the presence of the current monarch. The first queen to emerge usually destroys those not yet emerged, leaving her in charge. Often, however, she and her mother will remain in the colony, both going about their queenly duties—laying eggs. If the old queen is still producing some workers, the colony enjoys a burst in population. Eventually, the older queen expires, leaving her daughter the sole monarch.

You may find several supersedure cells on a frame and several frames with supersedure cells. Or, there may be only one or two supersedure cells in the whole colony. It depends on available resources and the availability of larvae of the right age for workers to raise as queens.

Drone-Laying Queen

Occasionally, you will have a queen that you have just purchased—or even one you have had for some time—that lays mostly unfertilized eggs, which produce drones. This can happen if a queen was not mated, or was poorly mated because the queen producer did not have enough drones to mate with the queens, or if the weather during the queen's short window of opportunity for mating did not allow her to fly to drone congregation areas.

She will appear normal, and if she is accepted by the colony, she will begin laying eggs. She lays them in regular worker cells, but none of them are fertilized; therefore, they all produce drones. This is very confusing for the colony, and also for you, initially.

It will take seven to ten days to recognize the situation, which is a great waste of time for the colony because this queen needs to be replaced immediately. This imbalance can also happen when an older queen eventually depletes her store of sperm and is unable to produce fertilized worker eggs. This is usually first noticed in the brood area. The usual solid pattern of closed cells will have open cells in places and a few larger, bullet-shaped drone cells scattered throughout the center of the frame, instead of along the edges, as is the normal location of drone cells. It occurs gradually, over two or three weeks, so you should notice the increase and order a replacement queen. The colony, too, usually recognizes this condition as a failing queen because the population becomes unbalanced, signaling a series of behaviors resulting in supersedure. To prevent a lapse in laying or a battle among emerging queens, look for supersedure cells and remove them before introducing a new queen that you have ordered.

For a variety of reasons unmated or poorly mated queens have become more common, especially when produced in early spring. Not enough drones to mate with, drones that are damaged by *Varroa* or have been exposed to *Varroa* control chemicals in their hives, and inclement weather so queens can't fly all contribute to this increase. This has become epidemic and queen acceptance or longevity has plummeted in the last decade. You need to be keenly aware of this possibility and know what to look for, and be ready to act if it happens in your colony. This means that the long-time rule of not interfering with a new colony for a week or so has to be modified. If you wait that long the colony will be hopelessly queenless if the original queen doesn't take. Give her no more than three days before you look for any brood, and what that brood is.

Unite a laying-worker nucleus colony, on top, with a strong, healthy nucleus colony on the bottom. Remove the excess paper from the outside and wait a few days. The bees will (almost always) sort this out themselves, leaving a single colony with one queen and no laying workers.

Laying-Worker Fix

A variety of mishaps—old age, injury, diseases, mites, exposure to pesticides—can befall a queen and cause her to stop laying eggs. Normally, a colony will note this change and begin preparations for producing a new queen: an emergency supersedure. Critical to this event is the presence of eggs, or larvae that are three days old or younger. These very young larvae have had only royal jelly as a diet. Older larvae have been switched to the less nutritious worker jelly, causing permanent physiological changes in their development. But sometimes communications break down, and the message that the existing queen is malfunctioning doesn't make it to the workers. By the time the problem is discovered, if at all, there may be no eligible larvae or eggs available, and the colony cannot by itself produce a new queen.

Without the queen's regulating presence, the ovaries of some workers begin to develop, and they gain the ability to lay eggs. Because they have no capacity to mate, all of the eggs they lay are unfertilized and will develop into small but functioning drones. The eggs are laid in regular worker cells.

Because a laying worker is smaller than a queen, many of the eggs she lays do not reach the bottom of the cell and cling to the sides. And because there are several, perhaps many laying workers, you may see several eggs in a cell. Other workers remove multiple eggs and raise a single drone larva in each cell. Initially, the overall pattern on a brood frame will be spotty, with some cells unoccupied, some with multiple eggs, some with normal-looking larva, and perhaps some capped drone cells. It is a confusing mess. Left to its own devices this colony is doomed, and if it's your colony, intervention is necessary.

It takes a colony several weeks to reach this sad state, and experienced beekeepers, rather than invest the time and effort required to save the colony, simply let it expire. This choice becomes obvious late in the season when even heroic efforts usually prove futile. You can expend extraordinary energy and cost to right this wrong, but a late season laying-worker colony is a lost cause.

However, if your laying-worker colony is discovered early in the process, or early enough in the season, there is a good chance it can be saved. Here's how to combine it with one of your other colonies. Reduce the laying-worker colony to one or at most two broodnest boxes. By now the colony is weak, so combine frames from the two or three brood boxes into one or two boxes. Put most of the brood and as many bees as possible into a single box. To do this, first remove any empty frames from the box into which you are going to put these frames. Then, remove the frames that have brood in them, along with the adhering bees, and fill the empty spaces. Take the rest of the frames and shake the bees into the new boxes.

Then, remove the cover, inner cover, and any honey supers on a nearby strong, healthy colony (with a queen). Place a sheet of newspaper over the top of the frames, and using your hive tool, make a slit in the paper between two or three frames, removing the excess from the edges. Place your laying-worker boxes directly on top of the newspaper, replace any honey supers above this, and close up the colony. This is called the *newspaper method* of uniting colonies. It is easy and generally successful.

The bees from both sides gradually remove the paper (in a day or several days) and carry it outside. In the process, the chemical messages from the queen below, coupled with the multitude of bees from the colony below that begin streaming up, essentially overwhelm the addition. The queen's pheromones spread throughout both colonies and begin to bring the two together. At the same time the laying workers are affected by the queen's pheromones and slow down or stop laying. They generally don't last very long anyway. After a week or so, the union is as complete as it is going to be, and where there were two, now there is one colony.

A similar technique, if you have several colonies to draw from, is to add brood and a new queen, but not at the same time. First, add frames of open brood from several colonies—three, four, or five if you have them—to the laying-worker colony, replacing drone-only brood frames. The overwhelming presence of young worker brood in the laying-worker colony has the same effect as introducing the entire colony to a queenright colony as above. Allow this overwhelming effect to take place for two or three days as the colony begins to care for all their new charges, and then introduce a new queen. Open a few of the drone cells on the discarded frame to check for *Varroa* mites, and then discard. If plastic foundation, simply scrape off the wax and reuse the

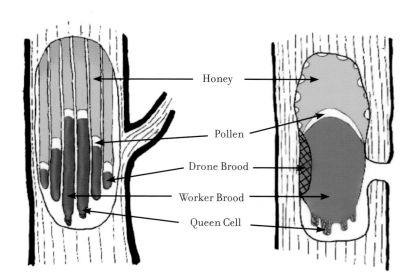

Two views of a typical hollow tree cavity showing the arrangement of honey, pollen, and broodnest. Queen cells are produced on the very bottom of the broodnest, where the wax is the newest.

Honey

Pollen

Drone Brood

Worker Brood

Queen Cell

The queen lays eggs in the broodnest. You can see them in the bottoms of these cells.

A typical broodnest frame has honey at the top and brood in the center. Between the brood and the honey is a narrow band of pollen.

frame. If wax, pop it in a freezer for a day and put it in a strong colony that will clean out all the dead drones. It's probably not a good idea to simply switch frames because of the *Varroa* in the drone frames.

Preventing this extreme measure and the resultant colony loss is, certainly, less work and less expensive. Colonies that are queenless almost always have some distinct, noticeable behaviors that provide clues to the situation.

Broodnest

A broodnest in a typical colony is oval shaped, or actually, shaped like an American football. This is where the queen deposits her eggs, the larvae are fed, and the sealed brood is kept. You have to imagine the broodnest in three dimensions. It is not a single frame, but rather like an odd-shaped sphere suspended within the confines of the square box you have put them in. Though the volume and location of this football-shaped sphere changes during the season as bees enlarge it in the spring and summer, then reduce it toward fall, until it is very small by early winter, the shape is relatively constant.

The cells in the top third or so of the broodnest have pollen placed in them by returning pollen-collecting foragers so the pollen is close to the young bees that need this food. Nurse bees, which need pollen in their diet to complete development, spend their first few days in the warm, safe center of the nest, close to the food and to the queen. They also need pollen to produce the glandular food they feed to the developing brood. Keeping pollen close to the broodnest is a matter of efficiency and necessity.

Surrounding the pollen ring, on the sides and above the broodnest, the bees store ripened honey. The honey stored closest to the brood is continuously replenished because it is constantly used for food. Above the broodnest, usually separated by a layer of honey used for food, the bees store the surplus honey needed to feed the colony during the part of the

year when plants are not producing nectar. The temperature in the central part of the broodnest, when brood is present, is held at a constant 95°F (35°C). The design of the broodnest serves several purposes. When the colony is young and small, the broodnest begins close to the top of the nest. As the nest expands, the broodnest area migrates toward the bottom where there is empty space, following the expansion.

At the end of the growing season, when brood rearing slows or ceases and nectar and pollen are no longer available, the living arrangement changes. Without a large brood area to protect and keep warm, the bees stay close to the larger mass of stored honey above and to the sides of the broodnest. They continue to move in the nest until they run out of food or until nectar and pollen are available again.

During this time, the queen may continue to lay eggs, but she tends to follow the cluster of bees as they move up the nest. If she ceases laying, she stays with the cluster. If the nonproductive season is not severe, the broodnest remains in its original location because the bees do not need to cluster to stay warm, or there are enough warm periods that food is easily obtained wherever it is stored in the hive.

Understanding this pattern of movement and how the bees construct their nest is important in managing a colony over the seasons. Anticipating what the bees are going to do allows you to prepare adequate space for them to move into. Replacing the combs of the broodnest after three or four years of use has become mandatory to provide the queen, brood, and young bees with new, clean wax and to make sure old wax is moved out. Old wax harbors many problems—agricultural and natural toxins returned to the nest by the bees, beekeeper-applied mite control chemicals absorbed by the wax, and spores of diseases affecting bees.

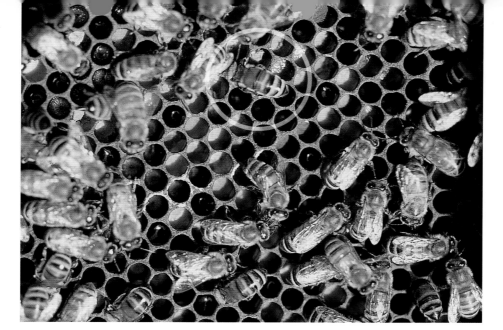

The worker bee that appears in the top center of this picture has her head in a cell, feeding the larva inside. Note the two bees in the bottom center of the photo. They are transferring nectar from a forager to a food-storing bee. This is part of the nectar-ripening process, which occurs in the broodnest on what is often referred to as the dance floor.

Worker bees are smaller than both queens and drones. They are by far the most numerous bees in a colony. The queen (circled) and workers are shown here for comparison.

The Workers

During the active season, a typical honey bee colony contains a single female queen, a few hundred drones, and thousands of female workers.

Workers raise the young, build the house, take care of the queen, guard the inhabitants, remove the dead, provide metabolic heat when it's cold and air-conditioning when it's hot, gather the food, and accumulate the reserves needed to survive the inactive season. When all goes well, workers also provide a surplus of honey for the beekeeper to harvest.

Queens and drones are fairly task-specific for their entire lives. What makes workers so interesting, and so complicated, is that their tasks change as they age, yet they remain relatively flexible and can switch between tasks when needed.

A worker starts as a fertilized egg, with half of her genetic traits taken from her mother, the queen, and the other half from one of the many drones with which her mother mated. She emerges from the egg as a larva, and for the following three days she is fed a diet identical to that of a queen larva. After that, her rations are cut (see *progressive provisioning* on page 57). As a result, her reproductive and some glandular organs do not fully develop. She is not as large as a queen, is incapable of producing queen substance, and is unable to mate. After three days as an egg, six days as a larva, and twelve days pupating, she finally emerges as a fully formed female adult worker honey bee. These times, of course, are not exact. They can be affected by the season, nutrition, external temperature, and the health of the colony—but for prediction purposes they are close enough.

Like any newborn, the first activity of this new worker is eating. Initially, she begs food from other bees in the broodnest area, but soon she begins to seek and find stored pollen. This high-protein diet allows her glands to mature for future duties. She stays close to the center of the broodnest—the warmest and safest part of the colony—which is also where most of the pollen is stored. Within a day or so, she joins the labor force, learning increasingly complicated tasks as she matures. She begins in the broodnest, removing debris from vacated pupa cells, pulling out the cocoon and frass (waste) that can be removed. Others follow her, polishing the sides and bottom of the cells with propolis in preparation for another egg.

After three days, the eggshell, or chorion, gradually tips over and dissolves and a tiny larva emerges. The larva is fed thousands of times a day by house bees. Like a queen larva, it floats in royal jelly. After three days or so, however, its diet is changed to worker jelly.

🐝 *A foraging honey bee trying to gain entrance to a colony other than her own will be inspected by attending guards and, if found to be in the wrong place, will be physically removed. Several guards may join the fray if required. This system is imperfect, however—if traffic is heavy, a foreign forager laden with nectar or pollen is usually allowed inside. The same thing happens when drones are returning from mating flights in late afternoon.*

After a few days, the glands in her head (the hypopharangeal and mandibular glands) are nearly mature, and the worker begins feeding older worker larvae a mixture of pollen and honey. As these glands mature, she can feed hatchlings royal jelly produced in those glands. She is also able to feed the existing queen this glandular food. At the same time, she can groom the queen, help remove her waste, and pick up after her. As she works, she picks up and distributes tiny amounts of queen pheromone throughout the colony, assuring all inhabitants that all is right in the world—or informing them that all is not right, that the queen may be failing, or even missing. This activity is all important in maintaining the status quo in a colony or in initiating a change that will again bring balance to the colony.

After a few days of cleaning, feeding, and eating, this worker begins to explore the rest of the nest, traveling farther and farther from the center. Soon she ventures near the hive's entrance and begins taking nectar loads from returning foragers—the first step in turning nectar into honey.

🐝 *Worker eggs in brood cells. To find worker eggs, start by looking at frames from the center of the broodnest. Carefully remove the frame, because the queen may be on it and you don't want to injure her. Hold the frame in front of you, with the sun shining over your shoulder, straight down into the bottoms of the cells. The eggs are small, white, standing straight up, and centered on the floor of the cell. They are just a bit smaller than a grain of white rice. Finding eggs takes practice, but it soon becomes second nature. One way you can see eggs is to hold the frame so sunlight is straight over the cell opening. Then, barely tilt the frame left/right, forward/backward, and the shadow of the egg will appear on the sides of the cell wall. Look for the moving shadow and the egg will appear.*

If there is sufficient room in the hive for nectar storage, the workers continue to take it from returning foragers. As room becomes scarce, however, the workers become reluctant to take it. Foragers with average- or less-than-average-quality nectar can be turned down as the number of bees visiting high-reward flowers increases. The same goes for pollen collection, which is influenced by the number of young needing to be fed.

Thus, if there is plenty of storage room and enough house bees available to take the returned booty, foraging increases. In fact, more and more foraging-aged bees begin to forage, and more and more receivers are recruited, leaving fewer house bees to clean and feed the young, as well as fewer guard bees to watch the hive. All of these factors spur the colony to collect nectar and pollen at an astonishing rate.

These older house bees also take on other tasks as needed, such as house-cleaning duties—removing dead bees, dead larvae, and debris, such as grass and leaves. (For more information, see "Foragers," on page 72.)

Are Bees Aggressive?

It's important to note that bees are not angry or mean or aggressive. They are, however, defensive. This may seem a small distinction when you or a neighbor is stung, but guard bees do not seek targets. Rather, in their limited way, they perceive threats to their nest and seek to defend that nest. Stinging is *defensive*, rather than *offensive*.

Making Wax

By the time a worker is about twelve days old, her wax glands have matured. These four pairs of glands are on the underside of her abdomen. Wax is squeezed out of the glands as a clear liquid. It cools rapidly and turns into a white, solid flake. The worker uses her legs to remove the wax flake, and then manipulates it with her mandibles to build the hive's architecture. She may add an enzyme-like compound to the wax if it becomes so solid it can no longer be easily molded. Pure beeswax is used to cover filled honey cells or to build new comb for storage. New beeswax is mixed with old beeswax and a bit of propolis, for strength, when covering brood cells and for use in building bridge comb.

When bees build new comb on a sheet of beeswax or beeswax-covered plastic foundation, they are said to be "drawing out" the comb. That is because they use the small amount of beeswax available to start the hexagonal cells, and then add to this foundation new wax that they produce in their wax glands. The result is a frame that looks like this, with all-white wax cells.

Honey Ripening

Nectar collected from flowers is roughly 80 percent water and 20 percent sugar, though these percentages can vary greatly. Though other sugars are present, sucrose, a twelve-carbon sugar molecule called a disaccharide, is the predominant sugar. The sugar content of nectar varies depending on the environment, the age of the flower, and other factors. During the flight home, the forager adds an enzyme called *invertase* to the nectar to begin the ripening process. Adding the enzyme changes the twelve-carbon sugar to two six-carbon sugars: glucose and fructose, which are not monosaccharides.

When a forager returns, she gives the nectar to a receiving house bee. This bee first adds additional invertase, and then finds a location in the hive where she can further tend to the drop she has accepted. If the rush of incoming nectar is hectic, such as during a heavy nectar flow in the busiest part of the day, she will hang the drop from the top of an empty cell or perhaps a cell with a small larva. The droplet will hang from the ceiling of the cell, exposed to the warm air of the colony until moved later.

Eventually, the nectar, which has been acted on by the enzyme and evaporation, is reduced to a mixture that is 18 to 19 percent water and just over 80 percent sugar, or what we call honey. Individual droplets are collected when ripe and placed in cells. When a cell is full, it is covered with a protective layer of new beeswax.

When nectar has been ripened into honey, it is stored in cells in the broodnest area or above the broodnest area in the surplus honey supers. When a cell is filled with honey, house bees cover the cell with new beeswax for protection.

Guards

After two or three weeks, a worker's flight muscles are finally developed and she begins orientation flights around the colony. Even before this, however, the glands and muscles of her sting mechanisms have matured, and she is fully capable of defending the nest. Therefore, she becomes a guard. In a large colony in midseason, the number of dedicated guards at any one time is relatively small—maybe 100 or so. However, if there is a large threat, thousands of bees can be recruited almost instantly. These new guards are temporarily unemployed foragers, older house bees, and resting guards. Actually, at any one time, there are many workers in a colony that are gainfully unemployed. They are resting, waiting for more nectar to return, have run out of young to feed, are foragers with no place to go, or have built all the comb there is room for in a colony. They are on call so to speak, so when a threatening incident occurs and some guard bee somewhere in the colony emits a tiny bit of alarm odor, there can suddenly be thousands of bees ready to respond to the call to arms in addition to the 100 or so on active duty.

Guards perform multiple tasks. They station themselves at the colony entrances and inspect any incoming bees. This inspection is odor based, because bees have a distinct and recognizable colony odor. If a forager returns to a colony that's not hers, she will be challenged at the door.

Other insects are also challenged if they try to enter. Yellow jackets, for instance, may try to help themselves to a colony's honey or look for a protein meal of honey bee larvae or adults. When this happens, the thief is met by several guard bees that wrestle and struggle with the intruder. They will bite and sting the intruder attempting to kill or drive it off.

Animals that try to steal from a colony are also rebuked. Skunks, bears, raccoons, mice, opossums, and even beekeepers will be challenged, threatened, and eventually attacked. When confronted by a large intruder, such as a beekeeper, some guards will engage in intimidating behavior before stinging.

They will fly at the intruder's face (they are attracted to the face because of the eyes and expelled breath) without stinging. This action can be annoying but—if the beekeeper wears a secure beekeeping veil—inconsequential. If such warnings fail to drive off the intruder, more guards will be attracted to the intruder. If the intruder's attack on the hive continues, the bees will sting.

You can often confuse these followers by walking into a stand of tall shrubs or brush, or stepping out of the line of sight of the colony for a moment—behind a building or into a shed or garage. The guards should quickly lose interest.

If you are still being harassed, keep your veil on until they head back home. Smoking these bees does little or no good in deterring their behavior because they are following you visually as well as by odor. If this behavior is common in your hive, it is a good idea to replace the queen that is producing these defensive bees with one that produces offspring that are less troublesome.

◀◖▶

You will see guard bees on the landing board in this defensive position, challenging any bee or beekeeper that dares to enter. Front legs up, head lowered, mandibles spread, wings extended, making her look as fierce and large as possible.

Smoke That Spot

When working a colony, you may inadvertently kill a bee by crushing her between hive parts or squeezing her with a finger. Even in death she may get revenge because the very act of dying causes her to release some alarm pheromone. Other bees will notice and respond. Smoking the colony masks the alarm pheromone, reducing the number of guards that respond to the alarm. If you are stung while working your colony, you are marked, but you can reduce the response of other guards by quickly removing the sting apparatus from your skin. Scrape or pull it out, then puff smoke on the sting site. This remedy helps reduce the attack but is a less-than-perfect solution.

When a honey bee stings, her sting pierces the skin of the intruder (or perhaps the cloth of the beekeeper's suit, or the leather of the beekeeper's gloves). The sting is a three-part apparatus, made of two barbed, moveable lancets and a grooved shaft. The lancets are manipulated by muscles. The shaft is connected to the organs that produce the venom and acid that are injected into the skin.

After the sting is embedded, these muscles alternately contract and relax, continuously, pushing the lancets deeper into the skin of the victim as the barbs hold the lancet and keep it from being removed, all the while the venom is flowing from the grooved shaft into the now-open wound. Each contraction pushes a barbed lancet farther into the skin with the venom gushing down the shaft going deeper and deeper.

Because the lancets are barbed, the bee cannot extract them. When she makes her escape and flies away, the sting apparatus and even some of her internal organs are torn away, remaining in the victim's skin. This is seldom a slow, methodical process. Guards approach an intruder, land, sting, and escape in usually less than a couple of seconds. You seldom see the bee that leaves her mark. The end result, though, is that when she leaves much of her internal organs behind, she is mortally wounded. She may, however, continue to harass the intruder. You may see one or more of these bees when you are working a colony, with entrails hanging from the end of their abdomens. They eventually perish, having died in defense of their home.

During a bee sting, the sting mechanism releases alarm pheromone while the muscles are pushing the lancets deeper into the skin. Alarm pheromone is extremely volatile, spreading rapidly in the air, and it alerts the whole colony that an intruder dangerous enough to sting is threatening it and serves as a call to arms. It marks the intruder, enabling other guards to home in on the sting site and further the attack. If the intrusion continues, the number of guards recruited increases until many, many reinforcements are in the air. This increase in guards usually drives off the intruder.

Guard bees make sure they are successful in thwarting your intrusion by continuing the attack as you leave the colony, or even the apiary. This behavior is variable, however. If there is a nectar flow occurring, with many bees coming and going, and the weather is cooperative, guards will seldom follow you farther than 12' (3.7 m) or so. However, the same guards may follow you much farther if there is a dearth of nectar or if the weather is cool and cloudy.

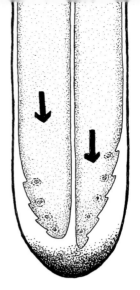

🐝 *A lancets barb shaft: The two lancets of a bee's stinger are barbed and work independently, but in unison, as they push deeper and deeper into the skin of the intruder. The shaft behind the lancets funnels venom into the wound that the lancets are producing.*

Foragers

When a worker matures and ventures outside the colony on a routine basis, she becomes a forager. This period of her life starts when she is about three to four weeks old, but may be sooner if the colony needs foragers. She may be a scout bee, seeking new sources of nectar, pollen, water, or propolis; then collecting some and returning to the hive to share her newly found information. Or, she may be recruited by another scout bee or forager to visit a productive patch of flowers.

If food is dispersed evenly, foragers exploit nearby areas, circling the colony. This is seldom the case, however, because flowering trees, shrubs, and weeds grow where they can, rather than where bees would prefer them. And as the season progresses nectar and pollen sources come and go. Therefore, the forage area changes from day to day. To add to this complexity, some plants produce flowers, nectar, and pollen during only part of the day. Cucumbers, for instance, bloom from very early in the morning until about noon. Bees visiting a cucumber patch learn the daily schedule.

As mentioned earlier, workers forage for nectar, pollen, water, or propolis, but not all at the same time. Some foragers gather nectar only and continue in that work for as long as the nectar is received back home. Some collect only pollen. Others, however, collect both during the same trip. We'll examine foraging for water and propolis later, but the relationship between bees and flowers is not only fascinating but critical to the success of the colony.

Robbing

Honey bee foragers have a fundamental goal—finding food. Most often that food is nectar or pollen from flowers. Other food sources can include floral food, sugar syrup in a feeder, or honey found in another hive. Small, weak colonies with few guards are unable to defend themselves from an onslaught of foragers from other hives, but they will try. Guards will fight to the death to keep strangers from stealing their hard-won stores, but if they are overwhelmed, the colony will be robbed of all of its honey, and in the process, many of the bees will be killed.

Beekeepers can inadvertently expose a weak colony to robbing by opening it during a dearth, or at any time foragers from other colonies are having difficulty finding food. Opening the colony sends an aromatic message—food!

There are several signs that a colony is being robbed, and it pays to recognize them before the colony is destroyed. There will be many bees at the entrance fighting, with workers balling up together with five, six, or maybe ten in a ball. There will be individuals rolling on the landing board and falling off, and all the time more and more bees arriving at the colony. Bees from one, two, maybe all the rest of the colonies in your yard can become involved.

Because of the mayhem and fighting, alarm pheromone fills the air. You may even smell the banana-like odor. Guards rush out of their colonies searching for the source of the alarm pheromone but will be unable to locate a typical intruder. When this happens, they can become defensive in a hurry, stinging everything and anything for several yards in all directions. A robbing situation can become fatal for the robbed colony and dangerous for people and pets in the area. If you suspect the colony you are working, or perhaps one you just finished, is being robbed, you have a responsibility to protect that colony before it succumbs. Immediately close all entrances tightly to all colonies in your yard using rags, sticks, or even handfuls of grass. Apply smoke to every colony to disorient the inhabitants and disrupt their rush to rob. Close up the colony that is being robbed, making sure that upper entrances are closed. Seal off the front door with grass. This stops outside bees from entering and allows the colony being robbed to reassemble its forces.

Honey robbing by outsider bees can begin during a beekeeper's inspection, when honey is exposed. If robbing begins, close up all colonies as fast as you can. Restrict the entrances of the colonies doing the robbing by inserting grass or a reducer, and completely close up the colony that is being robbed until the behavior ceases. A weak colony can be killed by robbing bees during the melee.

Bees that were robbing will continue to investigate previous openings, trying to gain entrance. If you've done your job, they won't be able to enter and in anywhere from a few minutes to a few hours will stop trying. When they return home, they won't be able to get in and will wait outside until the door is open. When it appears that life is back to normal, it will be safe to slightly open one entrance to the colonies. Leave the robbed colonies entrance reduced for days, or even weeks if the colony remains small or weak. Don't work any colony during a dearth, and know and watch for signs of robbing. It may be the worst thing that can happen to your colony and to your good reputation as a beekeeper.

Yellow sweet clover and its cousin white sweet clover are introduced species initially used as cattle food. They have escaped and are common nectar and pollen sources in much of the northern United States. However, invasives are severely disliked by naturalists, and eradication is on the agenda. Moreover, they are not nearly as good for cattle as other hay crops.

Finding food is the job of scout bees. Experienced scouts seek food using the color, shape, markings, and aroma of flowers. They learned that particular flower shapes, colors, or aromas signaled a reward and return often, or look for similar signs elsewhere. Beginners may recognize a familiar aroma, learned when they were in the hive, and investigate.

When a scout locates a promising source, she investigates its value. She lands on a flower, if it is large enough, or on a nearby stem or leaf so that she can reach the flower. She extends her tongue, called a *proboscis*, folding its three sections together to form a tube, and sucks in the nectar. She may scrabble in the center of the flower or brush against anthers in her pursuit of nectar, gathering pollen on her body hairs.

When full, she leaves the flower, circles the patch a few times—to get her bearings by noting landmarks and the position of the sun—then heads for home.

The Dance

Upon a forager's return from gathering nectar, she will be followed around the hive and touched and smelled by other bees. If the patch of flowers she visited was only so-so in its reward, the scout will half-heartedly recruit other foragers to visit the patch, but if it was outstanding, she will dance energetically to encourage others to visit it.

Because the forager has found a new patch, she usually tries to recruit other foragers to visit the patch. She initiates the dancing behavior on the comb in the lower part of the broodnest where other foragers gather, either waiting to be recruited or offloading pollen and nectar from a recent trip.

At the same time, the house bees who meet the scout on the dance floor have a message of their own. If there is a shortage of nectar in the colony and room for new stores, the house bees will unload the scout almost immediately. If, however, there is no room, house bees can refuse incoming nectar, effectively shutting down foraging activity. A critical situation confronts the colony at this point: If room is limited, yet there is a strong nectar flow on, they may decide to place nectar in the broodnest, where the queen is busy laying eggs. Doing so reduces the space in which the queen can lay eggs and may initiate swarming behavior if it is early spring, or cause them to only forage for what is eaten on a daily basis. Sometimes, and in extreme cases, this may lead to nest abandonment.

Other food-storing bees work with the pollen brought in by foragers, who dump it in a cell near the broodnest and leave it there, heading out for more. Young workers pack it into the cell, using their heads as rams, until the cell is nearly filled. They leave a shallow space at the top of the cell to be filled with honey, which acts as a preservative for long-term storage.

As hazardous as being a guard bee may seem, it doesn't hold a candle to the dangers encountered by a forager. When out in the field, a lone honey bee can fall prey to birds, spiders, preying mantises, and an array of other predators. The weather, too, works against her, with sudden showers, rapid temperature changes, or high winds making flying difficult. Other dangers include rapid automobile traffic and even flyswatters.

One danger that can threaten nest mates as well as the forager is insecticides. The forager can contact these poisonous chemicals in any manner of ways. She may be sprayed directly while in the field foraging on blooming plants. She may visit plants that have been sprayed some time ago and make contact with the residue. Depending on the pesticide and the how long ago it was applied, the forager may die very quickly and not return to the hive. With direct sprays, however, she often collects nectar, pollen, or both, then returns to the colony to slowly die there, while the cargo she brought back kills others in the colony. The result is usually a large mass of dead bees found inside and outside of the colony in a day or two. Depending on how much of the poison was encountered, the entire colony may perish. While once the most common form of honey bee poisoning, this occurs much less often now than in past decades.

Recent changes in pesticide delivery, however, have changed the way death is brought home. Many pesticides applied to crop plants now are systemic. That means the chemical may be sprayed directly, but more often it is applied to the seed when the crop is planted. It is absorbed by the plant as it grows and remains active for the entire life of the plant, killing anything daring to try to consume it. Of course the nectar and pollen are compromised with these chemicals and are brought back to the hive to be eaten or stored for later. The amounts of the chemicals in the nectar and pollen are not enough to kill outright, but rather deliver a sublethal dose again and again. Ultimately they reduce the ability of the colony to ward off other troubles, pests, and diseases, and enhance the damage caused by some of these other issues.

And what happens when honey bees come in contact with fungicides? For decades scientists considered this class of chemicals benign to honey bees because direct application to adult foragers did not seem to cause mortality. However, research has shown that when these chemicals were returned to the colony and fed to larvae, they caused damage, even death, as they affected the digestive organisms used by bees. The net effect was that brood was killed some time after the fungicide was applied. Later, other types of insect-controlling chemicals, such as growth regulators, were found to be causing similar problems.

It is difficult to completely avoid agricultural pesticides, even when you live in the city or suburbs, but be aware of them and what symptoms will appear if your bees encounter them.

If the forager avoids these dangers, old age will finally claim this five- or six-week-old bee. Foraging is the most personally expensive (excluding stinging) behavior of honey bees. The muscles deteriorate, body hairs are pulled out, and wing edges become frayed. One day, too tired or slow to make the flight back to the hive or to escape a predator's attack, her short, purposeful life ends. This is how it should be.

Communication

Something we don't often think about is that, with the exception of a very small area near the front door, the entirety of the hive's interior is pitch black. Any small cracks are sealed with propolis. Everything that goes on inside is done by touch,

feel, and smell. Yet, when foragers are out in our world they navigate by light and sight, by color and location.

Therefore, the returning foragers must translate their visual experiences in the outside world to their nest mates using nonvisual methods. When a forager discovers a new flower patch and has tested the quantity (to some degree) and the quality of its harvest, she returns to the nest to advertise its location and potential.

When she returns to the hive she goes to an area in the broodnest close to the main entrance. There, available foragers congregate. Also there are those bees who take nectar from incoming bees, and those very young bees that are cleaning cells and caring for the young.

To advertise the patch, the returning bee begins the famous waggle dance, which indicates with some—though not exact—precision the location of the source of her harvest. The information includes the distance (measured in expended energy during the trip home) and direction (where the sun is in relation to the colony and the patch). The value of her find is communicated to others by the intensity and duration of her dance. Unemployed foragers will come close to taste and smell the nectar, and some will follow her dance through several performances and eventually leave the hive in search of the source. There are many, many foragers recruiting simultaneously on this dance floor, but unemployed foragers do not sample each dance. Rather, they pick one and follow. They don't follow several, evaluating and comparing the differences.

The accuracy of the dance in pointing additional foragers to the patch is fairly reliable but not infallible. Factors such as obstacles (tall buildings) and head or tail winds enter into the equation as well. But outgoing foragers also have the scent of the floral source to help guide them, and a downwind approach can assist in locating the patch. Even so, many recruits leave the dance floor to find this floral patch, only to return in a short time to try again. They didn't get all the instructions, it seems.

Apple blossoms are nearly universal and provide much-needed nectar and pollen in the spring when they bloom.

Metamorphosis

Honey bees undergo what entomologists call complete metamorphosis. There are several ways insects grow. Complete metamorphosis describes the maturation of an insect from egg to adult. Because insects have hard exoskeletons, they cannot increase their size or the number of internal organs at will, so they produce a skin, grow into it, shed that skin, and produce a larger one. They will do this several times until they are as large as they should grow. Each of these stages is called an *instar*. Honey bees have five instars. In the last instar they cease feeding and produce a thin, silklike cocoon that covers the body. House bees, cued to the change when a larva quits eating, cover the cell with a mixture of both new and used wax. In twelve days, the transformation is complete, and a new adult pushes and chews her way free of her youthful confines. Her metamorphosis is complete.

(1) An egg is pictured standing on end, held there by glue used just for this purpose. In the cell to the right of the one containing the egg is a first instar larva, already floating in royal jelly fed to it by house bees.

(2) Larvae grow rapidly, going through five instars in six days. Here are two different instars (development stages between molts).

(3) When the larvae are ready to pupate, they stand upright in the cell, stop eating, void their digestive systems into the bottom of the cell, and prepare to spin their cocoons. Noting this change, house bees begin covering the cells with a mixture of beeswax and propolis. Now is the time that female Varroa mites enter the cells to parasitize the larvae. (Varroa mites are discussed in the next chapter.)

(4) Several stages of pupating workers are shown here. The wax cappings have been removed to show the developmental stages. At top left is a pupating worker nearing maturity, whose eyes have already developed color. At the center of the bottom row is a worker nearly mature enough to emerge as an adult.

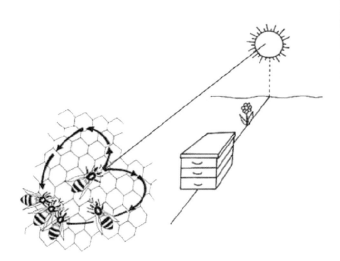

A simplified diagram of the waggle dance, which conveys information on the location of a food source in relation to the hive and the sun. The actual dances are much more complicated.

Every plant's pollen is distinct in shape, markings, color, and nutritional value. This is sweet clover pollen, greatly magnified.

Pollen loads are taken into the hive by the forager, who then puts her hind section into an empty cell near the broodnest or one partially full of pollen, and kicks off the pollen loads (often called pellets in beekeeping literature). The cells are left partially full; later honey will be added on top as a preservative.

Those that find the patch, however, will return and will recruit additional foragers if they found it to be profitable. This communication allows a colony to exploit many patches simultaneously. More rewarding patches will be highly recruited whereas less rewarding patches will be abandoned.

At the same time, the colony must adjust its capability to accommodate this influx of nectar. Returning foragers will actually recruit nonforagers to become food-storers by performing what is called the tremble dance.

Coordinating intake and storage as efficiently as possible allows a colony to quickly exploit as much of a good nectar source as possible. Moreover, it allows the colony to adapt to a changing environment to best exploit new sources, and it minimizes the time individual foragers spend searching. The greatest benefit is that not all of the colony's resources are focused on a particular flower patch. If a field of alfalfa that had been supplying the groceries for a week or more is suddenly mowed, only some of the foragers will be affected because others have been actively chasing the sunflowers, clover, and other sources nearby that were blooming at the same time.

Pollen—Pure Flower Power

Pollen is produced in a flower's anthers as part of the reproductive process. When mature, the anthers dehisce, shedding their pollen. Individual pollen grains are somehow transferred to the stigma of a receptive flower, which, depending on the species, can be the same flower, different flowers on the same plant, or flowers on different plants. Pollen travels into the female part of the flower—the ovary— and produces the seed and the endosperm surrounding it. (Think of apple seeds and all the rest of the apple surrounding

When pollen is abundant, a colony will often store nearly entire frames of it. An overwintering colony can use two to three full frames of pollen to feed brood in the spring when brood rearing starts, before pollen-bearing plants are blooming. It is said that it takes a full cell of honey, a full cell of pollen, and a full cell of water to raise a bee to adulthood, so you can see how much of these elements are required to produce a full-size colony of 60,000 or more bees in a season.

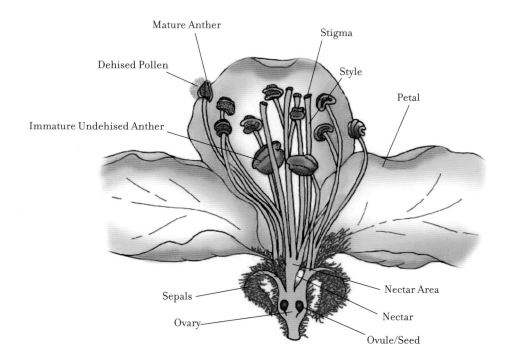

Mature Anther

Dehised Pollen

Immature Undehised Anther

Stigma

Style

Petal

Sepals

Ovary

Nectar Area

Nectar

Ovule/Seed

This illustrates the parts of a flower, showing an anther dehiscing (shedding) pollen, and the rest of the floral parts necessary for reproduction.

the seeds as the endosperm.) Pollen transfer is accomplished by wind, moths, butterflies, bats, birds, and a great variety of nectar- or pollen-feeding insects, including honey bees.

In their quest for nectar, honey bees come into contact with pollen on a routine basis, because flowers don't produce nectar until the pollen is mature. Some plants, such as cucumbers, produce male flowers that have both pollen and nectar but no ovaries for seed production, and female flowers that have nectar only and can produce seeds. To accomplish pollination, a honey bee visits both flowers (floral fidelity), receiving nectar from both and pollen from the male flowers. She then shares that pollen with the female flower during her visit.

Pollen grains have a minute negative charge, and bees have a minute positive charge and thousands of multibranched hairs capable of attracting, capturing, and holding pollen grains. Foragers clean most of these pollen grains out of their hair using their legs and carry them home packed in the corbiculae, or pollen baskets, on the outside of the hind legs. But while the bee visits other flowers, some pollen is transferred, and the plant has accomplished its goal.

Pollen is the only source of protein, starch, fat, vitamins, and minerals in a colony's diet. By weight, pollen has more protein than beef, and is the best food for developing larvae and young adults, and for producing brood food.

A colony will collect nectar and pollen from thousands of plants daily, and from hundreds of different plant species during the course of a season. This diversity provides a balance of the essential nutrients from pollen needed to grow healthy larvae and maturing house bees, and a rainbow of stored pollen inside the hive.

Collected pollen is returned to the hive by the forager, who promptly dumps it into a cell that already contains some pollen. House bees pack the pollen into the cell tightly so space is carefully used. Pollen is stored near the broodnest, where it is used at an amazing rate. Sometimes, an extraordinary amount of pollen will be collected, and entire frames will be filled with this multicolored food. Stored pollen fills only a part of a cell and is covered with honey for preservation and later use.

Water

Water is crucial to honey bee survival. Nectar, which is mostly water, provides some of the needed moisture in a colony but not nearly all of it, and certainly not enough.

Foragers will collect water when there is a need in a colony. Water is used to dissolve crystallized honey, to dilute honey when producing larval food, for evaporative cooling during warm weather, and for a cool drink on a hot day. A full-size colony at the height of the season on a warm day will use, on average, a half gallon (2 L) to a gallon (4 L) of water in excess of the nectar collected, daily.

Water isn't stored, like nectar or pollen, but rather is added to honey, or placed in cells or on top bars so it evaporates, cooling the colony in the process.

Foragers seem to seek water sources that are scented, and one of the greatest scented water sources in an urban or suburban neighborhood is a chlorinated swimming pool. The rising aroma is like a neon sign that says "Free Water Here— All You Can Drink." Stagnant pools of water and swamp land are also attractive because of the scent, and roadside ditches with rain runoff will see bees, as will storm drains and farm animal yards. The latter are especially attractive because of the minerals and nutrients that are dissolved in the water.

Foragers will mark unscented sources of water with their Nasonov pheromone so others can locate the source too. Once bees learn of a close, accessible, and reliable water source they will remain loyal to it. If that source dries up they will find another source and visit that until it, too, goes dry. It is in your best interests to keep their daily drink in your backyard.

Try the following techniques:

- Set an outside faucet to drip on a slanted board all the time during flying season.
- Install a water garden, or a bird bath, that automatically fills when water falls below a preset level.
- Invest in an automatic pet- or livestock-watering device that refills itself when the container falls below a certain level. Check occasionally to make sure animals haven't overturned or upset it, but this should be a nearly flawless way to provide water, without constant inspection, all season long. Be certain to shut it off when it gets cold, if it gets cold where you are.
- Rooftops can be troublesome as running water isn't normally available (unless you are fortunate enough to be on one of the increasingly popular green roofs in cities). As a result, you have to be doubly attentive to providing water

Foragers collect water where they find it, when they need it. Ponds like this one are ideal because they have an odor and are easy for other foragers to find.

AT ALL TIMES! Fill several pails of water at a time to make sure they don't run out. Have rain catchers available to take advantage of rainfalls, and provide floats or edges to containers to accommodate the bees. Some cities, New York for instance, can and will fine you if you do not have water nearby.

- If you provide a standing pool of water, say a child's swimming pool or a large barrel or pail, scent the water so it is initially easy for bees to find. Buy one of the many syrup additives made from essential oils.
- Put in your own swimming pool (admittedly, a significant investment).
- If none of these are possible, you can provide water on a regular basis using a Boardman feeder in the front door of the colony. It is easily visible so you will know when it is running low, but it will need to be filled on a regular basis. Forgetting is not an option because your bees will find water, and it will probably be somewhere you do not want them to be.

Propolis

Honey bee foragers collect nectar, pollen, water, and propolis—a wonderfully mysterious substance. Most plants have evolved some form of self-protection from microbial, insect, or even animal predation. Thorns, stinging hairs, bitter flavor, and poisons all are used by plants to thwart being eaten. A technique used by some species is the exuding of a sticky, microbially active resinous substance that covers leaf or flower buds while they are developing to protect them during this tender stage. Other plants secrete resinous substances around wounds for the same protection.

Bee foragers collect this substance by scraping it off buds or wounds with their mandibles and packing it away in their pollen baskets. Because freshly collected propolis is sticky, other bees help remove the mass when the forager returns to the hive.

Propolis varies in color, odor, flavor, and other microbial activity, depending on the plant from which it was collected, ranging from nearly black to brown, red, or gold. Also, the various microbial chemicals vary, as do the other compounds. In spite of this variability, however, no matter where it is collected, propolis is more similar than different, suggesting that bees seek plant resins with a detectable quality. The protective aspects of propolis should not be underestimated: It is active against potentially pathogenic bacteria and fungi in a hive.

Bees use propolis to seal cracks and crevices in the hive that are smaller than a bee. This makes this space unavailable to other creatures that may cause harm to the bees or the honey. If small hive beetles are present in your area, bees may actually build prisons of propolis on top bars and other locations to contain the adults and keep them from laying eggs on frames.

These cracks and crevices include the spaces where the boxes are joined together, where the inner cover and the top box meet, and where frames rest on the ledge in each box. If propolis builds up on this ledge, bee space is violated. Every so often, you must remove all frames and clean off the propolis from each box. When removing frames, clean the propolis off the lugs of the frames to prevent buildup.

The bees also use propolis as plaster to smooth rough spots on box sides, frame pieces, bottom board edges, and so on. Sometimes they will seal the entrance almost completely, leaving only a bee space doorway for them to use.

The propolis that is placed between boxes, or between the inner cover and the top box, will be sticky and gooey (at best) on warm days, but it becomes brittle when the weather turns cold. Even when it is cool, separating boxes or lifting the inner cover will cause the seal made by the propolis to *snap* loudly, alerting the bees that an intruder is present. It pays to keep those edges clean.

Harvesting Propolis

This goo can be profitable. It is a raw material used in homemade treatments for scratches, sore throats, and other minor ailments. Some propolis varieties from South America are said to be very good antibiotics and are marketed as such.

If harvesting propolis is an option for you there are companies that will purchase your collections for good money. Beekeeping supply companies sell propolis traps so you can collect propolis on an industrial scale. Installed like queen excluders, the mat is a mesh of slits, all smaller than bee space that the bees will fill with propolis to seal. When the mesh is full, remove the mat, freeze it for a few days, then bang it on something solid that is covered with newspaper, and much of the propolis will be dislodged, ready for collection. Then, reapply the mat and collect again. You can also collect the scrapings from frames, sides, doors, and wherever you find it in your hives. Do not mix propolis with beeswax because it will darken the wax, and will cause beeswax candles to burn poorly.

Propolis will be a part of your beekeeping life. For honey bees it is essential. For beekeepers it will be irksome, ordinary, and for some profitable.

Removing Unwanted Propolis

Always have a bottle of ethyl alcohol in your tool kit to remove propolis from your hands, hive tool, and smoker. Use a rag for the equipment, but just pour some on your hands and rub the propolis off. When you get propolis on your clothes (and you will), before washing use a commercial spot remover full strength and let it soak overnight. Once washed, propolis stains are permanent.

Honey bees use propolis to stick things together to keep them water- and windproof. A cover, such as this one, can be difficult to remove when it's warm outside because the propolis will be sticky. When it's cold outside, the propolis will be brittle—when you open the cover, it will vibrate and snap, an activity the bees inside frown on.

When routinely working your colony, take a moment to clean the frame rests and frame ends of sticky propolis. If left too long, the buildup will cause the frame tops to violate bee space. Also, clean frames help keep your hands and clothes clean when working the colony.

When propolis is warm, it has the consistency of chewing gum and will stick to anything it touches.

Honey bees will cover objects too large to move in the hive—such as these two mice—with propolis. The propolis retards bacterial and fungal growth and keeps the bodies from rotting. They just dry out when covered with this material.

Drones

Drones are the males in a honey bee colony. As such, they are different from workers and the queen in their physical makeup, their activities, and their contributions to the colony.

A normal colony will produce and support a small number of drones during the growing season. In a full-sized colony at midseason, as many as 1,000 drones, in all the stages of development, may be present. Drones produce their own pheromones that are recognized as part of the general aroma of the colony.

Drone honey bees are produced from an unfertilized egg laid by the queen, which means that the genes carried by the drones come from only the queen. The cells in which drones are raised are a bit larger than worker cells and, like worker cells, are part of the comb on a frame rather than hanging below it or butting between combs, where queen cells are constructed.

Drone cells are almost always located along the edge of the broodnest area, often in the corners of a frame, and the cooler part of the broodnest. This placement helps drone development; drone larvae and pupa do best in temperatures just a degree or two cooler than the 95°F (35°C) in the very center of the worker brood area. Drones take twenty-four days or so to develop from an egg to an emerged adult. They spend six and a half days, plus or minus, as a larva, and are fed a diet that is a bit more nutritious than a worker's diet but not nearly as rich as a queen's. Workers, by comparison, are larvae for about six days and pupa for approximately twelve days. Queens are larvae for five and a half days and pupae for only seven and a half days. A colony invests a lot of food and energy in raising drones.

A drone larva sheds its skin as it grows (as do worker and queen larvae). When the process is complete, the workers cap the cell with a mix of old and new wax, propolis, and other material. Because of the drone's large size, these cappings are not nearly flat, like a worker's, but domed to provide additional room, and are often referred to as bullet-shaped caps.

After twenty-four days (depending on the broodnest temperature, this time frame may vary by a day or so), an adult drone emerges. They are vastly different in appearance and function from their worker half-sisters. Drones have no sting apparatus. (A sting is part of a worker's underdeveloped reproductive system.) They are larger than workers, have comparatively huge eyes that reach to the tops of their heads, and have a stout, blunt abdomen.

For the first two days or so, they are fed by workers; then, while they learn to feed themselves, they beg enriched food from workers and begin to eat stored honey. After a week or

An adult drone can be identified by its large size, and fuzzy, blunt-tipped, stocky abdomen. Note the large eyes, extending all the way to the top of the head. The wings are about as long as the abdomen, unlike the queen's wings, which are only half as long as the abdomen.

so, they start orientation flights near the colony, learning landmarks and developing flight muscles. When weather permits, they begin mating flights. Drones do not mate with queens from their own hive. They fly to drone congregation areas (DCAs) in open fields, open spots in woody areas, or at the edges of large woody areas. Undisturbed areas serve as DCAs year after year for future generations. Drones tend to be fairly indiscriminate when looking for queens in the drone congregation area (DCA). They will chase nearly anything in the air above of the mating area, such as a stone thrown into the air.

Because drones cannot produce wax, cannot forage, and cannot clean house or guard the hive entrance, they are expensive for a colony to support. That investment continues to be borne by a colony to ensure that the genes of the drones' mother—the queen—are carried on in the general population of honey bees. There comes a time, however, when that price is too high for the hive to pay. If, during the season, a dearth occurs and food income is limited or nonexistent, the colony will, in a sense, downsize its population. They preserve worker larvae the longest and remove the oldest drone larvae from the nest first. They simply pull them out and literally eat them outright, conserving the protein, or carry them outside. If the shortage continues, they remove younger and younger drone larvae. At the same time, the queen stops producing drones to reduce the population even further, and eventually, if the shortage becomes desperate, they will eliminate any remaining adult drones.

At the end of the season, the colony no longer invests in drones at all. The queen ceases laying drone eggs, (for more information on "drone-laying queen" see page 63), and the workers forcibly expel all or most of the adult drones. Outside the nest they starve or die of exposure. Their retirement program isn't very good.

Seasonal Changes

To begin, imagine three generalized temperature regions: cold, moderate, and warm. The warmest areas are semi- to near tropical; warm regions are moderate, with winter temperatures hovering around 50°F (10°C) and a bit colder at night; and cold regions have winters falling to -20°F (-29°C) or colder.

In late winter, when the days begin to lengthen, the queen begins or increases her egg-laying rate. In warm-winter areas, Italian queens slowed down over winter but probably didn't cease laying eggs. Carniolans, Caucasians, and Russians probably did stop.

Worker bees consume stored pollen and honey to produce both royal and worker jelly to feed the brood during this time since there are no incoming provisions. As the population increases, so does the need for food. Within a month or less in warm areas, early nectar and pollen sources become available to supplement stored food.

In the moderate regions, it's still cold in late winter, but not so cold that managed colonies can't be examined by beekeepers on the occasional warm day. It'll be a while before good flying weather arrives, but brood rearing is increasing rapidly, and the need for stored food is critical.

By early spring, in the warm and moderate regions, the population is expanding rapidly as the early food sources become plentiful and the weather is increasingly favorable for foraging. In the northern hemisphere, dandelions begin to show in warm areas in February, and populations approach the critical stage. Additional room for expanding brood and food becomes a limiting factor, and swarming behaviors can be observed. Two to three months after the days being to lengthen, swarms emerge in the warmest areas. In the northern hemisphere, this happens in mid-April to late May, and late May through June in the colder areas, though these times vary depending on the location, management procedures, and weather. Don't set your management practices by the calendar but by what the colony is doing. The weather sets the stage, and the beekeeper has to be ready for swarming, food storage, queen replacement, and pest and disease control.

Food is abundant in the spring in most warm regions, but as late spring and early summer approach, resources often diminish, and by mid-July they are mostly gone. From then until early fall there is often little forage available, and colonies go on hold, living on stored food (if enough was made during the flowering season, and the beekeeper was not greedy at harvest time).

Here, a queen, worker, and drone are pictured. Note that the queen, shown at top, has a long, tapered abdomen. Surrounding the queen are typical workers, which are smaller and have striped abdomens. In the center is a drone. He is larger than the workers. Note the distinctive large eyes.

In moderate and cold areas, nectar and pollen sources come on strong right after the spring swarming season, and given adequate storage, a colony will collect most of the season's surplus crop starting in mid-June in moderate areas, and early July in the cold areas north of the equator.

In many places in the moderate and cold regions, there may be a slow nectar- and pollen-producing period in midsummer. Commonly called a *dearth*, this situation can last from a couple of weeks to a couple of months. A prolonged drought is another matter entirely and needs to be addressed as an emergency rather than a seasonal change. If food stores are all used, feeding may be necessary. Depending on location, this period is often called the *June gap* or the *July gap*—a short time when little pollen and nectar is available.

By late summer in the moderate to cold areas, fall-blooming plants kick in, and often there is another short but intense collection time. The duration of this collection time, called the *fall flow*, may last for a month or a bit longer, depending on how soon rainy weather and early frosts set in. It's an unpredictable time. Warm and tropical areas seldom have this late summer/early fall flow, but may if rains are abundant. Pay attention.

🐝 *On a typical brood frame in the broodnest, you can find both worker and drone brood. The worker brood is in the center of the frame, which is the warmest part of the nest. Drone brood is ordinarily along the cooler edges of the frame, especially as shown here, circled in the bottom-left corner. Drone brood is capped with large, dome-shaped cappings, which make them easy to spot on a frame. They can sometimes be found on top bars in the broodnest when brood frames are crowded.*

By late November, everywhere north of the equator, most plants have finished flowering, and the colony begins its slow time. Where winter is extreme, the bees cluster in the colony, work to keep warm using metabolic heat generated by vibrating their wing muscles and living on the honey and pollen stored during the past season. There may be extended periods when the bees can't break cluster and move around in the colony because of extreme cold. The occasional warm day enables them to move around inside the hive to get close to stored food and to take cleansing flights. The cluster of bees stays warm in a hive, but the air inside the colony is pretty much the same as outside. It's cold in there.

In warm regions, temperature extremes don't exist or don't last very long, and the bees can generally move and fly most of the time, though there may be little or no forage available in winter because it is dry, or rainy.

In some areas, such as the American Southwest, the bounty of spring depends to a great extent on winter rains. An average to abundant winter rainy spell means early spring plants will respond with exceptional flowering and lots of nectar and pollen to harvest. After that, the heat of summer dries up most nectar sources. Late summer rain may bring on a fall flow that holds the colony over until a spring flow begins again. Of course, in all regions agricultural crops can break the rules of nature. Irrigation allows plants to bloom when and where none would normally, providing nectar and pollen sources when they would normally not exist.

As stated, this is a generalized overview—the big picture—of the seasonal cycle of a colony. Plus, we've looked at how a honey bee colony responds to the world, and we've even looked at the individuals in the colony—workers, drones, and the queen—and have a good feel for what to expect from each of them. We need only one more piece of the puzzle—you.

There is so much to wonder at when working with bees that honey and beeswax are, and should be, only one of the rewards.

Winter Cluster

In the winter, honey bees require protection from the cold temperature and wind. During summer, they work to seal all the cracks in the hive so the interior is relatively windless. Beekeeper-provided wind breaks (evergreens or fences) help as well. Inside, the bees do the rest.

Roughly in the lower center (top to bottom, sides to sides) of a four-box, medium eight-frame hive is the broodnest portion of the colony. Figure the top half of the bottom box, almost all of the next two boxes side to side and top to bottom, and the bottom quarter of the top box as the broodnest portion, with the rest of the top box full of honey. For the most severe winter locations the fourth, top box should be all honey. There should be the equivalent of three frames of honey, both sides in the bottom box and the center two boxes on the outsides, and the equivalent of six frames or more of honey in the top box. The rest of the space should be mostly covered with bees, and some brood, which will be mainly in the second and third box. Brood, however, will vary depending on location and type of bee. When the outside temperature drops below 57°F (14°C), the bees congregate in the broodnest area. They crawl into some of the empty cells and fill the spaces between combs. Separated by thin wax walls, this fairly compact mass of bees is called the winter cluster.

The winter cluster is like a ball of bees—the outside layer of the ball is composed of bees tightly packed together and acting as an insulating layer, with their heads facing toward the center of the mass. Nearer the center, the concentration of bees isn't quite as dense, so interior bees can move a bit to get food or care for any brood present. This is where the queen spends the winter.

The temperature required to raise brood is about 95°F (35°C), so the bees on the outside of the cluster vibrate their wing muscles to generate heat, which warms the bees in the center. The bees on the bottom of this cluster work harder than those above because their heat rises and warms the center bees and those above the center. Those at the top mostly act as insulation. At the very edge of the cluster the temperature reaches but does not fall below 45°F (7°C).

Obviously, the bees on the very outside of this cluster cannot sustain themselves for long at that temperature, so they gradually move toward the center, and the warm, well-fed bees from the inside move toward the outside to replace them.

As the outside temperature falls further below 57°F (14°C), the cluster begins to reduce its size, shrinking uniformly. This action reduces the surface area of the ball, and the bees in the insulating layer move even closer together, reducing heat loss from the interior and forming dead air spaces between their bodies, using the hairs on their bodies to capture that warm air. As the temperature drops more, the ball continues to shrink. This configuration can be maintained for a short time, but eventually the bees in center will have consumed all of the honey they can reach.

Bees do not warm the entire inside volume of the colony. The only warmth generated is kept in the cluster and is not wasted on empty space. This concentration of warmth is efficient from a heat-conservation perspective, but there is a downside. When the outside temperature becomes very cold, the cluster cannot move to reach more food. If the temperature remains very cold for a long time (less than about 20°F (- 7°C), the bees may starve when the food inside the cluster is gone and there are not enough bees to reach additional food, even when food is only inches away. Having enough bees to cover enough comb is important for this reason.

A colony may die for many reasons, but it will starve during the winter for only two reasons: There isn't enough food, or there aren't enough bees. Both of these conditions are preventable.

Take location into account: A colony in the southernmost quarter of the United States will be able to fly all year long, even if rain and cool weather hamper foraging some days. The next quarter or so north will have several periods where flight is restricted and stored food will be required, so having about 50 pounds (23 kg) of stored honey on Thanksgiving Day is a good practice. That's about the equivalent of eight or nine frames, both sides in the upper three boxes. For the next quarter north, winter can be severe at times, for extended periods ranging from several days to perhaps three or four weeks in some years. For this general area, about 75 pounds (34 kg) is recommended. For those colonies in the far north, 100 pounds (45 kg) of stored honey is the rule, and a fourth box completely full of honey will accommodate that.

And how many bees are enough? In the southern half or so of the United States, winter weather is seldom severe enough long enough that a cluster can't move every few days to reach more food. In the northern half, though, it can be. Enough bees in a cluster will enable that cluster to reach the tops and the sides of the frames they are on so some bees can reach the other side and go above or even below the frame if necessary. Once some bees reach new honey, even if the brood is in the center of the frame and most of the bees need to stay there to keep it warm, food can be passed to distant bees, just like nectar is passed from forager to house bee. Enough bees allow expansion past the center brood to the edges, both sides, below, and above the cluster.

When the outside temperature warms, however, the outer insulating layer of bees expands, and the volume of the cluster expands with it, moving to frames with stored honey on the sides of the broodnest or above it.

When bees visit blossoms, they pick up pollen and carry it back to the hive.

If you can manage a garden in your backyard, keeping bees follows a similar routine. You need a commitment in time, but it's not a crippling amount of time. It's doing the right thing the right way at the right time that leads to success. And if you get it mostly right, most of the time, you'll enjoy the bees much more than if you are always behind. Just like your garden.

Be aware that much of what you will be doing, you will be doing alone, at least when it comes to management. Harvesting and bottling, even candle and soapmaking often receive help from interested people in your home, but working bees is probably going to be your single-handed task.

What's Outside

If you've been gardening for a while, you are aware of how your growing season progresses. This is a distinct advantage when you start keeping bees because you are already familiar with the usual benchmarks. Generally speaking, the differences between seasons depend on how close you are to the equator, no matter which side of it you live on. The closer you are, the more the seasons blend; the farther away you are, the more distinct they become. At the equator, the climate is tropical. As you move farther away it becomes semitropical, then moderate, and finally polar. In the northern hemisphere,

spring begins toward the end of March, and summer toward the end of June. Conversely, in the southern hemisphere, spring begins toward the end of September, and summer starts not quite the end of December.

Equally important is having a general idea of what the plant world is doing during each of these seasons. Knowing and managing colony activities in normal, rainy, or dry seasons, nectar or pollen dearth, and regular seasons, gives you a head start on bee management. Just as with your garden, you should learn when to plant, when to water, and when to expect to harvest.

This then is the macroenvironment in which you live. But closer to home, where your bees go, is a microenvironment that's equally important. Specifically, what plants are growing close enough that your bees can visit them? When do they bloom? The climate and the weather will affect, to a degree, when those plants bloom—let's say apple blossoms—by a couple of days, or as much as a week earlier or later in some years. All plant growth is dependent on day length, temperature, and available water and nutrients. In beekeeping literature there are a multitude of resources available that review those plants that are nectar and pollen producers, where they grow, and when they bloom throughout their growing range.

Nasonov Pheromone

When working a colony, you will often see worker bees on the landing board, facing the entrance. Look closely and you'll notice that their abdomens are raised in the air with the tip bent down just a bit. This position separates the last two abdominal segments, exposing a bit of whitish integument below. Located at that spot is the Nasonov gland, which produces Nasonov pheromone. Exposing that gland allows some of the pheromone to waft away. To help distribute this sweet-smelling chemical, the bee will rapidly beat her wings. She is said to be fanning or scenting. This is intriguing behavior, and only workers can do it. Broadly speaking, this is an orientation signal produced to guide disoriented or lost workers back to the hive.

Interestingly, when one bee begins fanning, it stimulates nearby bees to do likewise, and those bees that return begin to fan also. Very quickly you'll see many, many fanning bees on the landing board or the top edge of an open super, guiding their lost nest mates home.

This pheromone is also part of the glue that keeps a swarm together and all going in the same direction as it leaves its nest when heading for a new home.

Workers use this pheromone in a variety of other ways inside and outside the hive. You may see bees scenting at a source of fresh water. What you'll notice most, though, is that when you open a colony, the natural upward ventilation, partially driven by the body heat of thousands of bees, wafts up the commingled, subtle aromas of curing honey, stored pollen, and a good bit of Nasonov pheromone. This cocktail produces the distinct smell of a beehive. This is what makes all colonies smell mostly alike, but all a bit different. There is no other aroma quite like it or quite as attractive, to both a honey bee and a beekeeper.

When a hive cover is removed, some bees will fly away or may fall to the ground when the cover is placed on the ground. If they are unskilled flyers or too young to have flown orientation flights, they may become lost almost immediately because they don't know landmarks. Skilled or experienced bees that fly off will immediately return and begin scenting behavior by exposing their Nasonov glands and fanning their wings to drive the pheromone away from the colony. The inexperienced bees will pick up on the pheromone's aroma and follow it home.

This swarm has left its colony but has not yet found a new home. You can see bees scenting with their Nasonov glands exposed. Other bees are dancing, trying to convince the swarm that they have located a good home site.

Review and Preparation

Here's a brief review of what has been presented. Let's go over getting started, the equipment you'll need, the bees, and how to handle them throughout the season.

Before You Begin

- First thing, find out the rules and regulations that exist about bees and beekeeping where you live, or where you are going to keep your bees if not in your backyard.

- Check out all this with your family. Don't assume everybody is going to be as excited about bees in the backyard as you are, even if gardening and growing fruit trees are already a way of life. Be realistic about what you need in terms of yard space required, flight paths, equipment storage (and construction, if you go that route), cost, harvesting, and even beeswax activities if you decide to make candles, creams, or lotions. These will require room, time, and resources.

- Make sure you and your family members aren't in that tiny minority of the population that has honey bee allergies. But don't confuse normal bee sting reactions with troublesome, but far less likely, allergic reactions. If you are unsure, get everybody checked out by a physician so that there is no doubt. Be safe.

- Next, find out what your neighbors think of your newfound hobby. Even if there are no, or only limited, restrictions on having bees in your yard, if a neighbor has an irrational fear of honey bees (and certainly if they have a health issue), your plans may evaporate. You can pursue this, but living next to a hostile neighbor makes life difficult for everyone.

- Be realistic about your available time when deciding how many colonies to manage. Even if local regulations allow up to five colonies on a lot the size of yours, do you have the time to take good care of that many and the storage space for that much equipment? Begin with fewer hives, so that you can learn the ropes at a gentle pace.

- Learn as much as you can. Read books and magazines. Join the local beekeeper's association. Visit other beekeepers, and carefully use the Internet to see how other beekeepers do the things you'll need to do, not just read about them. Take a beginner's class in person if at all possible, but there are several excellent university-sponsored programs available online. Get all the catalogs.

Getting Ready

Set your calendar so that your bees arrive right about the time dandelions and fruit trees bloom where you live. The following preparations have to be done by then.

- Prepare the place the colony or colonies will be. Provide screens so that the colonies aren't visible to neighbors, or from the sidewalk. Make sure flight paths are up and away from where people spend time—decks, the garden, play areas, and especially neighbors' yards.

- Make sure your hive stands are tall enough to be out of reach of animal pests, and sturdy enough to hold one or more colonies weighing up to 150 pounds (70 kg) each, by season's end, and that the area around them is cleared and weed free.

- Provide a permanent source of water that will not dry up if a natural source is not available.

- Order your bees from contacts made at meetings or classes, or from suppliers, four or five months in advance, if possible, to ensure availability and timely delivery. Ask local beekeepers what race of bee works best where you live, and why they like them. Overwintering, honey production, pest resistance, and gentleness are all traits to consider.

Equipment Dos and Don'ts

- If you are considering a beginner's kit, look carefully at what it includes and doesn't include, and the price and quality of the items. You get what you pay for. Price each item individually to see which kit is the best for the money. Be especially aware of the protective gear and the feeders. Customize your operation so that it fits you—not what somebody else chooses for you.

- Order your equipment early enough that you have ample time to prepare it all. Find a supplier that you are comfortable with, one that has all of the equipment you need and is willing to answer your questions. When purchasing additional pieces, remember that not all manufacturers' equipment matches—stick with a single supplier, at least until you are more familiar with beekeeping.

- For convenience, safety, and ease of use, I recommend starting with eight-frame, preassembled equipment, using screened bottom boards with tray inserts, and wooden frames with black plastic, beeswax-coated foundation. Either flat or decorative covers are effective, but the gabled covers are heavy and will not work as a surface to place additional supers on while examining a colony. Slanted landing board stands are completely unnecessary. Queen excluders are optional, but try one. A 1-gallon (4 L) plastic pails or hive top feeders are needed for feeding.

- Paint or stain the exterior-only parts of your equipment with neutral or natural colors. White stands out, brown and gray don't.
- Get a full bee suit (best) or jacket-style suit with attached, zippered veil (these are not available with beginner's kits, so watch for that). Start with two hive tools, the large-size stainless steel smoker with heat shield and the least bulky, most sensitive gloves you can find.
- Decide what crop you will seek—liquid honey or comb honey sections or cut comb honey.

Beekeeping Preparations

- Keep good records, and refer to them each time before examining your bees.
- Have a plan when opening a colony—know what you want to do, what you should be finding, and what maladies, if any, may be present.
- Become familiar with how a colony of bees operates. While learning, examine your colonies often, checking the brood-nest boxes for signs of a healthy, productive queen, healthy brood and the right amount of it, plenty of pollen and honey stores, the right mix of workers and drones for the time of year, and any maladies that are present.
- Manage your colony for swarm prevention and control, but know that some colonies will swarm, no matter what you do. Have a bait hive in your yard, if possible. Answer swarm calls with care and caution.
- Requeen every colony as necessary with queens from producers actively seeking genetic resistance to pests and diseases in their bees. This is the only long-term solution to reducing pesticide use in your hives. Ask hard questions, demand good answers, and expect to pay top dollar. Recall that bad queens never get better, and cheap queens are just that. Be aware that queens are more fragile now than they were twenty years ago, and losing a queen shortly after installation is a frequent occurrence. You need to know what to do when that happens.
- Be patient when introducing queens. Leave the cage in the colony at least five days before removing the cork, and then allow three more days for the bees to remove the candy. It's the safest way there is. And even that isn't foolproof.
- Monitor pest populations all season and be prepared to act if they suddenly increase. Treat if required, but don't treat "just because," or "by the calendar."
- Actively follow integrated pest management (IPM) techniques and methods to avoid or reduce pest populations and hard and even soft chemical controls.
- Always remember that honey is a food, and your family will eat it. Treat it as such.

Before you continue to the next section, here are a few guidelines to keep in mind. They're not carved in stone, but maybe they should be.

- There isn't a lot of work to keeping bees, but what you have to do, you have to do on time.
- Always dress and work with your bees so that you feel safe and secure, and your bees are not threatening anyone nearby.
- Your family, neighbors, and friends should not have to change their lives because you did (unless they want to).
- You should care for your bees the best you can, just as you do your other pets, recognizing your responsibility for their well-being, providing food, water, health, safety, and shelter.
- Beekeeping should not overwhelm your life. It is part of your life, not all of it.
- Sometimes you will be hot while working with bees.
- When it's not fun anymore, it's time to hang up the hive tool. If that's the case, plan an exit strategy by selling or otherwise disposing of your colonies and the bees. Don't abandon them, leaving the equipment to become a source of disease or pests and an attractive nuisance. Be a good neighbor.
- And the last guideline has to do with food. Enough good food all season long is the cheapest insurance and the best medicine your bees can get. It is your responsibility to see that they do.

CHAPTER 3 ❖About Beekeeping❖

This chapter is going to work from the ground up, starting like you do—with a brand-new package in your brand-new equipment. We'll shepherd that colony through its first year and into the next season, so next year you'll have all the information you need to continue.

Start by reading the section on installing a package. Get familiar with the sequence of events before you begin. Installing a package is a simple process and difficult to bungle. However, it can be a bit nerve-racking the first time.

Lighting Your Smoker

Your smoker, hive tool, and protective gear are all tied for first place as need-to-have beekeeping tools. But you also have to light and keep your smoker lit. If it goes out when you take your colony apart, you'll need to retreat to relight. Meanwhile, your colony stands open and unsecured.

Lighting a smoker is similar to starting a campfire or a fire in a fireplace. Start with rapidly burning, easily combustible tinder. Newspaper is perfect and readily available. Once it is burning, add less-flammable material and establish that. Then, if needed, finish with longer-burning fuel that will sustain the fire for as long as needed. The following nine-step process works every time.

🐝 *Step 2: Crumple the newspaper into a loose ball that doesn't quite fit into the smoker and light the bottom.*

🐝 *Step 3: Let the paper catch fire and the flame begin to move up without it reaching your hand. Push the paper to the bottom of the smoker using your hive tool, and puff the bellows gently two or three times to keep fresh air moving past the burning paper. It will flame up to or just over the edge of the top of the chamber.*

🐝 *Step 1: Assemble your tools: paper, pine needles (my fuel of choice), punk wood, matches, and smoker. Smoker fuel depends to a degree on where you live. If long-needled pines are numerous, your fuel issue is solved. But people use wood shavings, cotton waste sold by bee supply companies, wooden pellets, untreated burlap or twine, sawdust, or almost anything that's combustible and does not contain pesticides or petroleum products.*

🐝 *Step 4: Puff two or three times more until most of the paper is burning—but not yet nearly consumed—and add a pinch of pine needles or other fuel to the top. Don't stuff them in, just let them fall in.*

🐝 Step 5: Puff a couple more times until the flames from the paper reach up and catch the needles afire. Once they are burning well, push the burning needles down into the chamber with your hive tool, puffing slowly so air moves through the system.

🐝 Step 7: When the second, or perhaps third, batch of needles begins to smolder from the bottom, push the fuel down first with your hive tool, then with your clenched fist so the cylinder is about half full of compressed fuel. Keep pumping. The coals should be far enough below the top that you will not be uncomfortable. Push fairly hard until the fuel no longer collapses. Then add the more durable fuel, if needed, on top of the needles. Keep puffing slowly to keep air in the system.

🐝 Step 8: When lots of smoke rises when you puff, and if the fire doesn't quit when you don't puff for a minute or so, close the smoker, still puffing occasionally.

🐝 Step 6: When the first small batch of needles flames up, add a bigger pinch of pine needles, loosely. Puff several times so you don't smother the fire. Keep air moving through the system. This is when most lighting attempts fail because too much fuel was added, so no air is available from the top and not enough is coming up from the bottom. The smoldering needles starve for oxygen and the fire dies.

🐝 Step 9: When lit, the smoker should smolder unattended for up to a half hour or more. If it sits idle for a while without use, puff rapidly a couple of times so the smoldering coals flare up a bit, producing lots of cool, white smoke to waft over the bees.

Equipment Checklist for Installing Packages

Be prepared: You will always need some of these things, but almost never will you need all of them. Unless your tool storage is less than 10' (3 m) from your bees, missing something you need can be anything from a minor inconvenience to a major headache. It's always better to have a tool and not need it, than to need it and have to go somewhere and get it.

- Pliers (regular and needle-nose)

- Hive tools (two)

- Smoker and fuel (matches or lighter)

- Misting bottle filled with sugar syrup (1:1 ratio, sugar–water)

- Bee suit and veil

- Duct tape

- Gloves

- Extra box(es) to cover the feeder(s)

- Feeder(s): pails or hive top (if hive top, bring a pail of sugar syrup to fill it; if feeder pails, make sure they are full)

- Wire (a large paper clip works), or rubber bands (<10" [25.4 cm]) (to suspend/hold the queen cage in the new colony)

- Brick or stone (to weigh down the new colony's cover)

- Entrance reducer (must downsize the entrance to 1" [2.5 cm])

- A means to secure the entrance reducer (push pins, a small nail, even duct tape)

- Hammer and a few nails

- Toothpicks

- Notebook

- Pollen supplement patties

- Camera (This is a once in a lifetime event!)

- Bees (Please, don't forget them.)

When you finish using the smoker, empty its contents on a fireproof surface, making sure the coals are out. Sometimes when you dump out the fuel in your smoker, it will burst into flame because of the sudden availability of oxygen. This can catch you or other flammable material on fire very quickly. Use caution. You can also lay the smoker, on its side; without a draft from bottom to top, it will quickly go out. Still lit or still very hot smokers can cause fires if placed too close to flammable material. Make sure it's empty and cool before putting it away.

Occasionally check the intake air tube on the bottom to make sure it is clear, and scrape out the inside of the funnel. Accumulated ash and creosote will slowly close the opening.

If sparks from the fire are coming out of the spout, check your fuel because it may be nearly gone. If the fuel is still plentiful, grab a handful of grass and put it on the top of the fuel inside to stop the sparks.

Never aim your smoker at someone and puff. Besides causing the inhalation of smoke and limiting sight, flying sparks may ignite clothing. Also, use smoke sparingly on your bees. A little bit goes a long, long way in controlling the bees.

Package Management

Almost all beekeepers get their first bees from a business that sells packages. Bees are sold by the pound, and the most common size has 3 pounds (1.4 kg) of bees. With about 3,500 bees in a pound, a 3-pound (1.4 kg) package (the most common and most recommended) contains roughly 10,000 bees. Almost all are workers who have been removed from a

 A typical 3-pound (1.4 kg) package of honey bees. The bees hang around the queen, who is suspended inside the package next to the tin can that holds sugar syrup. The package has a cardboard or plywood cover on top, keeping the can, the queen, and the bees in place. A few dead bees will be on the bottom of the cage; if there is a ½" (1.3 cm)-thick or more layer, contact the supplier.

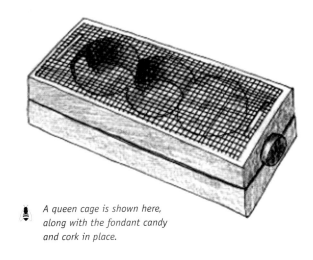

A queen cage is shown here, along with the fondant candy and cork in place.

Use a mister bottle to put sugar syrup on the screen so the bees have food. Use a new bottle, and don't soak the bees when misting.

single colony and placed in the package. Then, a queen not from that colony, in her cage, is added to the package, a feeder can put in its slot, and the top covered.

You'll get your package from a local supplier or through the mail. Order it far enough in advance to assure there are packages available for delivery when you want them. A good rule of thumb is to have your bees arrive when deciduous fruit trees or dandelions are about to bloom—April or May is about right. Your equipment should be ready before the bees arrive.

If possible, when ordering your bees by mail, make arrangements with the supplier to have them arrive midweek (they will probably insist on that so they don't sit in a post office over a weekend). Weekend mailings can sometimes be neglected in large post offices. Make sure your phone number (home, work, or cell phone for daytime messages)

Local suppliers drive to the package producer and pick up packages directly, reducing the delivery time and the stress on the bees. You can get packages either with a queen or without one and install the one you want later. Bees are a commodity, but the queen is the future of your colony. Don't settle for just any queen.

is prominently displayed on top of the package, with instructions to call you when the package arrives at your local post office. Remember to inform the post office that you must retrieve the package immediately. Some post offices will store it in an out-of-the-way place, such as a loading dock, or storage room. Unfortunately, these places can be too warm, which can be lethal to the bees if held there too long.

If you purchased the bees from a local supplier, you'll pick them up on the assigned day—usually a weekend—right after they arrive. In either case, get the bees home as soon as possible. These bees will be stressed—they have been several days away from their home colony with limited food, a strange queen, temperature extremes, and jostling.

Of course, you should get the bees into their new home as soon as you can. That may be just an hour or two, or a day or two, depending on your schedule and the weather. Cool, rainy weather is tolerable, but very cold temperatures—below 40°F (5°C)—are probably colder than you want, so wait a day. While waiting, keep your package in a cool, dark place, such as the basement or garage. Put down a couple sheets of newspaper to keep the floor clean.

You'll also need to feed the bees. Get a new, unused spray bottle (such as those used for misting plants), and mix up a sugar-water solution of one-third sugar to two-thirds water by volume, that is, 1 cup (235 ml) white sugar mixed with 2 cups (475 ml) warm water. As soon as possible, spray some of this solution through the screen directly onto the bees. Don't saturate them, but moisten as many bees as possible through both sides of the package. Do this at least a couple of times a day as long as they are in storage. The feeder could be plugged or empty, so don't assume they have food available.

In They Go!

You've prepared your family and your neighbors for your adventure long before this day. The site has been chosen and prepared, and your equipment has arrived and is ready.

If time works in your favor, plan to install your package in late afternoon or early evening. This timing helps your bees remain calm and assists in settling in for the night.

To begin, use your spray bottle to give the package a feeding. Then, take all your equipment out to the site and get it set up. Put the bottom board on your hive stand and one super with frames intact on top of that. Leave the next two supers, without frames, stacked alongside, and one super with frames intact. (You need four supers.) Don't forget the inner and outer covers. Prepare the sugar syrup for the feeder pail with a two-to-one water-and-sugar solution. Fill the pail to the very top, and make sure the lid is secure. Refill your spray bottle at the same time.

If all your equipment is new, your frames will have only foundation, either wax or wax-coated plastic. If you have combs from other equipment, place those in the center of the eight frames in the box. (Be certain, however, that the colony from which these frames came was disease free.)

Put on your protective gear (for more on protective gear, see page 41), light your smoker (you probably won't need this, but be prepared), make sure you have your hive tool and pliers, and bring your package to the hive site. If you want, bring a board large enough to fit securely on your hive stand to serve as a solid, dry working surface, but only if the hive stand is long enough to hold both the colony and the board, with a little room to spare. Otherwise, place the board on the ground close to the back of the colony's entrance.

Remove the cover, inner cover, and six frames from the eight-frame box that sits on the bottom board. Set the cover, inner cover, and frames behind the box or to the side. If you are installing more than one package, prepare all the boxes at the same time.

Position yourself behind the colony into which you plan to install the bees. Make sure you have all your tools—smoker, hive tool, and feeder. Set the package on the board, and using your hive tool, remove the cover. You may also need pliers for this. Remove any protruding nails or staples, and keep the cover close at hand because you will need it again in a few moments.

Take a look at the opening underneath the cover. You'll see the top of the feeder can, flush with the surface of the package top. There will also be a slot cut in the top with a metal strip in it. This strip is fastened to the queen's cage, which is suspended below in the mass of bees. It also allows you to easily remove the queen's cage. The strip should be long enough to grasp easily and to hook over the top of a frame.

Lightly mist the bees again. Using the corner of your hive tool, lift up the feeder can. Some are easy to catch and lift, but some will be a bit below the surface and more difficult to grasp. If you simply can't catch it, try lifting just a bit and grasping with the pliers. Hold it with the pliers and grab with your other hand.

Next, still holding the can in one hand, lift the package and thump it on the board so all the bees clinging to the feed can let go and fall to the bottom. Don't worry—they're covered in sugar syrup and don't care one bit. Try lifting and moving the package to make sure you have a secure grip. When comfortable, lift the package 1' (30.5 cm) or so and thump again. Lift out the can, set it on the board, and slide out the queen's cage—without dropping her. Cover the hole in the package with the original cover. Check to make sure the queen is alive and moving, and then put her cage into your pocket to make sure she stays warm.

Check to make sure the frames have been removed and that you have placed the wooden entrance reducer in the box into which you are going to dump the bees. Make sure the box that is going to go on top of the bottom box is close by, along with the inner cover that will go on the new colony.

When you're ready, you're going to thump the package again, then remove the cover and pour the bees into the cavity in the space created in the box when you removed frames. Pour as many as you can—shake them a little, but not much—and set the package, still containing a few bees, in front of the colony. This action will put some bees in the air, but don't be concerned. They are homeless, very confused, and not likely to cause you harm.

Carefully lower the frames you removed into the bottom box. Let them slowly sink as the bees on the bottom are gently moved aside to make room for the frames. Here's where queen introduction to the package can go in two directions. In the photos you'll see you can place the queen in her cage in the bottom box, held between two frames so the bees have easy

(1) Make sure you have a good hold on the can before thumping the cage. If not, it may fall off the support below, if there is one.

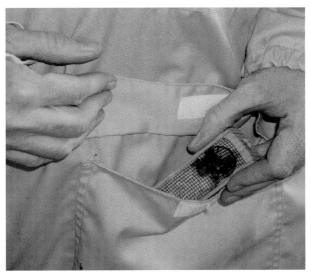

(3) When the queen's cage is free of worker bees, put it in your pocket to keep her warm and for safe keeping.

(2) Remove the queen's cage carefully. Don't drop her, if you can help it. Don't let the bees unnerve you. Gently shake or blow them off the cage.

(4) Quickly and carefully dump the bees into the space created by the missing frames. Shake the bees a bit to get them to come out, but don't worry about every last bee. You'll get them later.

(continued)

🐝 *(5) If your queen cage comes equipped with a hook or wire, loop it over a top bar, off center, and hang the cage so the screen faces the opening between frames. Consider securing it to the top bar with duct tape or a staple. You can't be too careful with a bug as expensive as a queen. The tape will make sure your queen stays where she should. If there isn't a wire or tab, and often there isn't one that works for installing in a package, use rubber bands.*

🐝 *(6) If you are lucky and have drawn comb in the frames in the box above the box you dump the bees into, install the queen in the upper of two boxes. The queen's cage should be placed between frames in the top box, with the screen facing down, and the cage actually resting on the lips of the bottom bars. Gently squeeze the two frames together so the queen cage becomes just a bit embedded in the wax comb on each side. This will securely hold the cage in place until she is released. This won't be possible if this is your first colony and you don't have drawn comb already, but it is very useful if you do.*

🐝 *(7) Place the feeder pail over the hole in the inner cover. You'll need two medium supers to enclose the pail.*

access to her. She can also be placed in the box directly above the bottom box, again held in place. Be sure you don't place her in the center, in case your feeder pail leaks. The flood would drown her.

Three criteria for inserting a queen cage are: easy access to the queen cage by the hive bees, being securely held, and out of the way of the feeder (if it leaks). Even the best of queens, however, in the best of cages, introduced the best way possible by the best beekeeper there is, has about an even chance of being accepted. This acceptance rate goes up if you are using local or survivor queens, or queens you have raised yourself, or queens the colony has raised, or you bought a nuc and the queen was already accepted. With new packages be prepared for queen loss and replacement.

Mostly this has to do with the miticides queen producers use to control *Varroa*. They are death on queens and drones, and what you get is the chemical fallout from their colonies—drones that aren't capable of mating, and queens that aren't capable of producing many eggs so are soon replaced by the colony's bees.

The short term answer here is that the more care you take in introducing your new queens, more food, and less interaction with bees they aren't familiar with, the greater the probability they will be accepted.

If you place her in a second box between frames, hold the frames secure, and place this box on the box with all the workers. Place a pollen substitute patty directly on the top bars to one side of the inner cover hole before replacing the inner cover, or putting on the hive top feeder. Leave the paper on so it doesn't dry out. Replace the inner cover, position the feeder if you are using a pail over the inner cover hole, and cover with extra supers. (Or, put on the hive top feeder, the inner cover, and close the colony.)

And you're done. Make sure the small entrance to the reducer is open to let the remaining bees not yet in the cage go in at their leisure (usually overnight is fine). Pick up all the extra stuff and tools you brought to the site and call it a day.

If you're still uncertain, reread this section so the sequence is clear in your mind. It's fairly simple and straightforward, but being familiar with it always helps.

First Inspections

There are many recommendations on how often to inspect your colony, especially a new package. (Checking the amount of available sugar syrup doesn't count as a visit.)

Oops! Did You Drop the Queen's Cage?

"I dropped the queen's cage! Help!" This happens, especially when you're wearing gloves. You have to get it. Here's how: Quickly put the cover back on the package. Look for the queen cage in the mass of bees inside. If you don't see her, gently roll the package until you spot the cage. If you don't already have a glove on, this may be a good time to get it. Gently thump the cage to confuse the bees and spray them with syrup. Remove the cover and reach in, grab the cage, and pull it out. Quickly replace the cover. It's that simple. If you can, don't wear your glove. The texture and feel of all those nonthreatening bees on your fingers and hand is a feeling like no other.

In the past, before the troubles with queens became endemic, it was thought that during installation you could simply remove the covering from the candy end of the queen cage when you were placing the cage in the colony and let the bees begin to release the queen immediately. This normally took two or three days—ample time for introductions to be made and relationships to be established, especially since the bees and the queen had already spent some time together in the cage during transit from there to you.

This is no longer the case; because of *Varroa* mites, and the chemicals used to combat them, the science and art of commercial queen production has changed. Now, there are fewer feral colonies to supply drones and as a result the queens today are sometimes poorly mated, or not mated at all. (This doesn't usually affect introduction, but if she isn't mated at all, she's still a virgin and will be treated as such by the bees.) Some of the chemicals used to combat mites in queen mating beehives have been shown to be a disaster on the health and well-being of both queens and drones, so they, too, have been producing less than stellar results. Queens exposed to certain chemicals die young, have trouble laying eggs, and are physically deformed. Then add in the fact that the varieties of queens are changing (Russians, hybrids, and so on) and the time needed for new queens and workers to become acclimated has become longer.

Do not remove the covering from the candy end of the queen cage for at least five days.

After the package has been installed, you must wait patiently. Watch the front door for activity, make notes, take pictures, watch more, and be patient. Check the food situation on top, but don't worry about the pollen patty yet. (See page 101 for more information on feeding pollen substitute patties.)

After five days, open the colony carefully, using as little smoke as possible. Peer between the frames and see what the bees are doing around the queen cage. Are there lots of them there, hanging on tightly? Or are they milling around,

This is comfortable behavior—the bees aren't clinging to the cage or biting the wire. Queen release now is safe.

When it is time, remove the cork on the candy end of the cage, using the corner of your hive tool. Don't remove the cork on the other end or the queen will walk out prematurely.

🐝 *This is what you might see when the workers are still aggressive toward the queen. They cling tightly to the cage, biting the wires and are difficult to push away because they are trying to get at the new queen.*

🐝 *If you leave your feeder on and the bees come up to feed, they may construct comb that surrounds the feeder and is attached to the sides of the box. It must be cleaned up. Remove the cover and inner cover, smoke the bees, remove the feeder, and remove the beeswax. Save the wax (it makes good candles, lotions, or wax for coating plastic foundation). Clean it all out, remove and fill the feeder, and replace the box and cover—and don't leave it alone for so long again.*

moving over and around the cage? Then, remove the cage, either by lifting the frame with the cage banded to it, or simply lift the cage out using the hanger. See if she is alive and moving, then gently lay her on the side on the top bars and observe. Do bees rush to the cage, and grab it tightly? Do they walk around and over the cage, and then walk away? Do some stay, touching with antennae, feeding the queen? Or do they cover the screen and holes and look as if they are trying to smother the queen? Let this go on for a few minutes. Be patient, and do not use smoke.

If the behavior seems casual, not urgent, laid back, then you can be relatively certain that the queen and bees have accepted each other and things will go as well as things can go.

If, however, the bees are still clinging to the cage and the screen, if after being removed they return immediately, you can be fairly certain that they haven't yet accepted each other and they need more time. Give them a few more days. Check again, and if they've settled down (they almost always do), the queen-to-colony relationship has been established. Releasing the queen is the next step.

If the behavior seems friendly, proceed to the next step. Lightly smoke the colony in the area of the queen cage so the bees retreat. Remove the cage and remove the cork from the fondant candy end with the corner of your hive tool or a toothpick. Test the candy with the toothpick to see if it's hard. If it's soft, your work here is done. If it's hard, very gently poke the toothpick through the candy, making a small hole all the way through it. Just be very careful not to jab the queen when you break through. (If you're apprehensive about this, remove the cage from the colony and yourself from the situation to a spot where you feel comfortable and carefully push the toothpick through the candy.)

When complete, replace the cage in the same place, making sure the screen isn't covered by any new wax comb that may have been built up around the cage or frame parts. In fact, you'll often find stray slabs of beeswax comb, so carefully remove them with your fingers or hive tool first. Replace the top box carefully, check the feeder, refill it if necessary, and put everything back together.

The queen should be released by the bees in another three days, so plan on checking again. If she's not, and if the behavior of the workers is still protective rather than aggressive, pull back the screen and let her walk out, heading down between frames. Don't let her fly away. Close up the colony.

After her release, the queen should begin laying eggs in a few days, at least within a week. After that time, check for eggs. You'll find them in the center two or three frames, probably near the tops of the frames in the bottom box.

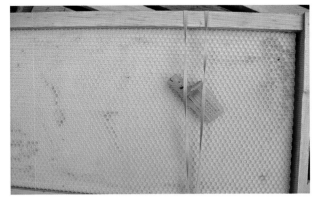

🐝 *If you are using only foundation and your queen cage doesn't have a wire, metal strip, or metal disk, use rubber bands to hold it to the frame. The candy end of the cage must be up, and as much of the cage's screen is exposed between frames as possible so the bees in the colony can reach, feed, and touch the queen in the cage.*

Look at all the frames that have some comb built on them if you don't find eggs in the center. Recall that eggs are tiny, standing straight up on end, and nearly the same color as the new wax—look carefully.

Once the queen is laying eggs you're over the first hurdle and don't need to inspect the broodnest for another ten to twelve days. Make certain the feeder stays full, however. The bees will continue to use the sugar syrup for some time, especially when the weather doesn't cooperate—a sure thing in the spring—and at night. For much of the first season, the colony is living hand-to-mouth with little chance to build reserves. The more you can help, the better off they will be.

So, what if you don't find eggs? It happens. Maybe she just wasn't mated at the producer's before she was sent, or the disease nosema affected her, or she was injured in transit or installation, or in spite of your best efforts and judgment she wasn't accepted by the colony and was killed? Look to see if she's there. In a colony this small, if she is alone and walking

Queen Replacement

If you lose a queen during introduction, immediately contact the business from which you purchased her and let them know what happened. They may offer a free replacement, but you pick up the shipping. Don't argue or debate their decision. They may have good advice on introduction techniques, or perhaps they have had mating problems and a replacement is due. Or you may have goofed. However, if they become defensive or obstinate (a rare occurrence), consider a different queen producer.

When the queen arrives, introduce her using the same techniques as before, mindful that this new queen didn't have two- to three-day acquaintance time in the package. She should be ready to be released after a week in the cage, so after five days or so, go in again and remove the cork and let the bees release her naturally, as long as their behavior was friendly after that time. After another ten or twelve days, you need to inspect the broodnest again to record the quality and quantity of the brood the queen is producing. There should be eggs and larvae present.

Also look at the sealed brood, if any, noting solid patterns in the center of the frame surrounded by open brood and eggs. Drone brood should not be present anywhere in the center. A drone-laying queen needs to be replaced immediately, as does a queen who has a lot of empty cells in her broodnest.

As difficult as it may seem, and as frustrating as it can be, never debate replacing a queen who isn't performing well. Bad queens never get better, and a bad queen will, at best, head only a mediocre colony and may cause its demise and death. Requeen at the first signs of a problem. It's a small investment in an entire season. Don't be cheap when it comes to queens—you will always be sorry.

Something to keep in mind about your package: No new adult bees have been produced for at least twenty-one days. Usually, it is closer to four weeks. During that time the bees that came in your package have been making beeswax comb, tending the queen, feeding the young, foraging, and guarding.

This three- to four-week period is about as stressful as it can get in a colony. The demand for food, especially pollen for the house bees to turn into worker jelly, is extreme. Sugar syrup is a wonderful carbohydrate source, but protein is needed too. Foragers are scrambling to collect pollen, especially because you are providing sugar. It's a hard time in your colony until the new bees begin to emerge. Your food, can mean the difference between success and failure in the first few weeks. But if you need to replace the queen, you don't have time to waste.

around she will be fairly easy to find. Listen for the telltale buzzing sounds of a queenless colony.

If you find your new queen, even if she isn't yet laying eggs, close up the colony and give her two more days to start laying. If after all this nothing's happening, something's wrong and she needs to be replaced. If you don't find her, order a replacement just as soon as you can—that day, if possible.

Honey Flow Time

A newly introduced queen starts laying eggs at a slow pace. She has already laid some eggs before the queen producer sent her to you, but then went for many days confined to her cage. Once released, this young queen's egg-laying rate builds slowly, starting at perhaps 100 eggs a day for a bit, gradually increasing to as many as 1,500 per day when all conditions are favorable. The rate depends on the health of the queen, the health of the colony, available pollen, favorable foraging weather, and adequate space.

The bees generally begin building comb in the center frames, using most of the space from top to bottom in the bottom box and anywhere from none to all of the frames from top to bottom in the second box from the bottom. Usually,

it's most of both, but the bottom of the frames in the top box and the top of the frames in the bottom box get built first. The ends of the frames are usually the last to be filled.

When there is comb being built on most of five or six frames in the bottom box, and four or five in the top box, it's time to add a third box. At the same time, switch positions of empty frames and frames with comb, but not brood, placing those with some comb next to the edge of the box and replacing them with empty frames, encouraging the bees to fill the now-centered frames next. This rearrangement encourages the bees to fill all combs, rather than use only the center frames in the boxes. If you don't switch frames, the bees may "chimney" their living quarters, building all the way to the top but only in the center. During this buildup time there may be a nectar flow from early and mid-spring blooms. Nonetheless, maintain the feeder at all times to ensure there is always sugar and water in the hive. Three days of bad weather could actually spell doom for the colony because they have very little stored food. They turn every ounce of carbohydrate they can collect, whether from your feeder or a nectar flow, into food and wax.

Here's a handy tip: When you add the third box for your brood, add the queen excluder and check and make sure you have two or three honey supers ready to add. If not, get them ready. When you add a queen excluder, move a couple of frames of comb without brood above the excluder so the bees know that going there is okay. Another tip: Paint the honey supers a different color than the brood box so you don't inadvertently use a honey super for a brood box or vice versa.

Feeding, like requeening, has evolved over the years and should be examined a bit more closely.

A package colony, especially if it starts out on foundation, starts out under a cloud. You are providing them with the carbohydrates they need, but everything they eat they turn into beeswax to get the home started, food for themselves, or food for baby bees once the queen starts to lay eggs. Feeding baby bees requires a lot of protein.

Also, no new bees arrive on the scene to help out for as long as four to six weeks. For those six weeks—the lifetime of a bee, by the way—the bees have to get acquainted with a new queen (and hope that she does her job, so they can do theirs), build a new home, collect food, feed babies, defend the nest, and take care of that new queen.

One solution: Feed, feed, feed sugar syrup (not high-fructose corn syrup) as a carbohydrate source as long as they will eat it. Feed pollen substitute patties for a protein source. (Over time, sugar syrup that sits on a colony will ferment, and the bees will

---

Tip:

Always clean your tools when you've finished inspecting a hive. Ideally you will do this between inspecting each hive in your apiary, but sometimes that's a bit unrealistic. However, definitely clean them when you are done for the day, or between checking different apiaries. Designate a small pail as your cleanup pail. Fill it half full of water and get a couple of metal scrubbies for doing dishes. To clean, scrub your hive tool, removing all bits of wax and propolis and honey. If your gloves are sticky, wash while still wearing them; it's only soap and water and it removes honey, propolis, and, most importantly, venom residue from any stings you received. Slosh the smoker with the water when done and scrub it off, too. Cleaning your tools nearly eliminates any chance of transferring disease from colony to colony or beeyard to beeyard. And there is nothing more distracting than having a hive tool so sticky from honey that you can't let go of it.

🐝 *Pollen substitute patties (available from bee suppliers) provide the minerals, proteins, and other nutrients honey bees need so they don't sacrifice the protein in their bodies to feed their young and have the resources needed to make food for the colony.*

not eat it, so monitor the feed source.) Another rule of thumb is to feed sugar syrup until they don't take it three times. Provide a pail of food, if they don't take it and it ferments, provide another, and then another. After the third you can figure they have enough. The same with the protein patty: If it dries out, provide another, then another. (A cautionary note: Watch for small hive beetle larvae consuming the patty. If so, remove patty and larvae immediately and do not add more food.) After the third try they probably are doing okay.

As before, purchase preassembled boxes and wooden frames with plastic foundation. Get them painted (or stained) right away and have them ready. It takes an hour or less for two or three supers. If you have to assemble them, plan on a

🐝 *Pollen you have collected in a pollen trap is a better food than a pollen substitute. You can purchase pollen that has been irradiated to eliminate any disease organisms. Pollen you collect can also be packaged and sold to supplement your income from honey sales (above).*

How Many Supers Should You Have?

If you are in a warmer climate, the nectar flow tends to be early, intense, and short, maybe only three to four weeks. Sometimes, it's less. Supers can fill fast, and room for incoming nectar can disappear almost overnight. Without room for nectar or honey, the rate of foraging will taper off. It may even stop. Or, the broodnest area may begin to fill up, reducing the queen's activities. For short, intense honey flows, put two supers on right away. This makes sure the queen isn't slowed and there's room to grow, especially if you miss a week's monitoring while on vacation or if you get busy in the garden. If the flow stops, the bees won't use the box and you can remove it unused.

There are mixed reviews on the best way to manage this situation, but if you restrict storage space and there is lots of incoming nectar, the bees will (almost always) begin to fill the broodnest, restricting egg laying. Recall the conditions associated with swarming behavior—restricted space, lots of bees in the hive, a queen whose egg laying has slowed, and a whole world to explore.

This is where choosing medium-depth, eight-frame equipment is truly a benefit. When it's time to add that additional super, first remove the partially filled box, then put the new box directly on top of the queen excluder that was added when the third super was put on. Your bees, accustomed to the excluder, easily move up to store nectar and honey, will continue to do so, filling the top super. On the way up they pass right through the new space and will begin to take advantage of it, as long as nectar continues to come into the colony.

lot more time.

After three to four weeks, depending on the weather, the third box should have three or four frames with comb and two or three with some brood. The broodnest should encompass one or two frames in the top box and most of the frames in the middle and lower box, with pollen and honey stored on the edges, and mostly or all honey on the very edge. The outer side of the frames on the edge may be barely filled with comb and a little honey this early in the season. Again, exchange frame locations to encourage the bees to fill the partial frames in all boxes.

Be patient—it may take twice as long as described to set up your colony. The key is to observe the sequence of events and the buildup of the whole colony. Weather, available protein, the building of new comb, and you, the beekeeper, all add stress to this new colony.

Adding a honey super provides the room necessary for colony expansion. But don't be caught short. Late spring nectar flows can be intense, and if your colony's population is strong, the right plants in bloom, and the weather suddenly turns warm and humid, that super can be filled in a week or ten days. It is astonishing how fast this can happen when everything works. Being slow on your part to provide room will reduce the honey crop at the end of the season, but that's not a negative if you're not sure what to do with hundreds of pounds (kg) of honey.

More likely though, nectar flows are slower, and the bees will take substantially longer to fill these supers, mostly due to erratic weather. Another factor that can be limiting is available forage. If your colony is near undeveloped land where a diversity of blooming plants exist, the nectar flow will speed and slow as nectar-producing plants flourish and fade. In a more developed area, nectar plants are primarily

domestic and the variety can be nearly infinite, but the quantity can be limited. You need to check on the rate of filling in your location, and add more space either as needed or as you want to plan your crop size.

When the first honey super has comb on five or six frames and honey being stored in four or five, even if it's not yet covered with wax, it's time to add a second super for additional space for nectar storage and, eventually, honey storage.

Keeping Records

If it hasn't occurred to you already, keeping a log of your colony's activities and progress is a good idea. Particularly for the first couple of seasons, making notes will force you to attend to the fundamentals, and the notes remain as a record of what happened when. If you have several colonies, some notes will serve for all, but each colony, you will learn, has its own distinct personality, requiring similar but different management actions. Even before you get your bees, decide how you will identify your colonies. "The third from the right" doesn't work after about a season. One good way is to simply number the colony with spray paint on the top of the cover. That way, no matter where the colony is, you will know all about its history. At the beginning of the season, record the following for each (now numbered) colony:

- Queen source and/or package source
- Location, position (north-, south-, east-, or west-facing) of the colony, and registration papers and inspection reports if you have them
- Condition of equipment

Then, include the following notes about each visit:

- Weather and time of day
- Date
- What's blooming—you need to look around a bit for this, but the major common plants are easy: dandelions, wildflowers, flowering shrubs. These major nectar producing plants bloom about the same time each year, but you need to learn when so you can prepare for honey flow time.
- Temperament of the colony—easy, busy, flighty, fussy, loud
- Depending on the time of year, note:
  - Queen cups present
  - Queen cells present
  - Spotty brood pattern
  - Number of eggs, open and sealed brood
  - Drones present
  - Honey and pollen present
  - Signs of pests or disease
  - Physical condition of combs and other equipment

Then note what activities you performed, such as:

Tip:

During stretches of bad weather, early in the season, when the bees' food sources are compromised, continue feeding sugar syrup until the third feeding jar has fermented without any feed being taken. It is the cheapest insurance you can buy for the health and well-being of your colony. Feed pollen substitute until they haven't eaten any for a month. Monitor the feeder and the patty. When they finally quit eating—no more sugar feeding and no more protein feeding—they have enough stored to last them through a bad spell, and they are comfortable with the natural resources coming in.

There is no substitute for keeping good records. Get a good notebook or a three-ring binder. Designate one page for every colony (spray paint a number on each colony, too). Or, create a page for every visit to the beeyard, and keep dated notes on each colony visited on each page.

- Requeened (note the source and breed) and color of mark—always get marked queens
- Fed, and how much
- Exchanged frame positions
- Removed old, black frames and added new frames. (And where did you get these, and what did they cost? It's not a bad idea to write on new equipment where and when you got it so you remember particularly bad or good suppliers.)
- Added brood or honey supers
- Couldn't find queen, or found a queen that shouldn't be there (your marked queen is gone)
- Applied medications (and when the next treatment is due)
- Harvested, and how much

Reviewing your notes before you next open your colony will remind you of actions you need to take, equipment you need to buy, problems you need to check, and what you should expect to find.

Keep records in a large notebook that is difficult to misplace. After a few seasons, your notes will be minimal because you will have mastered the routine management skills, noting only swarming dates, new queens added, medication applications if used, and the amount of harvest.

Opening a Colony

The first few times you examine a colony can be exciting, scary, and confusing. It's easy to get sidetracked and forget to do one of your planned tasks. So before you get started,

make a good mental note of why you are looking inside. This is good advice whenever you are going to examine your colony. Check your record book first, always. When you know why you're going in, you'll know what you need to do the job, so get everything together and bring it with you.

Before you begin, make doubly certain your smoker is burning well, but give it a reassuring puff every few minutes, just to be sure. If you're feeding, have your extra feeder full and ready. Bring supers or other equipment you may be adding, and have your hive tool in hand. It's not good to leave a colony open if you have to run back to the house.

This may seem obvious, but quickly scan the ground for rocks, branches, or toys. Stepping on one of these with your hands full of a box of bees can be disconcerting. Check to make sure nothing is on the hive stand where you plan to set your boxes, either. And don't put the things you brought with you in a place where they will be in the way before you use them. Whenever possible, minimize moving things more than you need to—it saves the back and the temper.

When everything's assembled, you're ready to puff the tiniest puff of smoke into the front door. This will offset the guards at the front immediately, reducing any flying. Then, step back and again look at the setting, making sure nothing's in the way and you have everything you need. This delay of a

To begin examining a colony, remove the cover of the colony carefully.

minute or so lets that little bit of smoke waft up into the lower box just a bit and contact the dance floor area, slowing that behavior also.

Remove the cover as carefully and as quietly as you can, but keep it close. If you have a telescoping cover, place it topside down next to the colony on the hive stand, about 6" to 8" (15.2 to 20.3 cm) from the colony, with the long side parallel to the bottom board. When you remove the inner cover, place it on the cover, turned about 30 degrees. Place any boxes you remove on the inner cover with the same orientation. This placement keeps bees from leaving the box from the bottom, and a bit of smoke puffed over the top will keep them inside. If there are boxes protecting your feeder, remove them and the feeder, putting them out of the way. Puff a half puff or so of smoke into the center hole before you remove it. Again, wait a moment or so before proceeding.

Pry up the inner cover with your hive tool, give a half puff, and let it down for another moment, then remove it completely and give a puff, maybe two if there are a lot of bees on the top bars and between the frames. Let them go down before you begin.

Loosen the closest or next-closest frame first. Use the curved end of your hive tool for leveraging and loosening frames.

Drifting

When you have more than one colony, some bees from one colony will drift from home to the other colony. If you have options on where colonies can go, which way they face, and how close together they can be, you can probably reduce drifting between colonies. Drifting is a problem when more bees drift one way than back, weakening the donor colony. Diseases and pests can ride along, too.

To reduce drifting, face colonies that are close together in different directions, with entrances at 90 degrees or 180 degrees to each other. Using boxes painted with different colors gives the bees a clue of where to go, as does a landmark, such as a bush.

If you find that one colony is collecting a lot of drifters from a nearby colony, you can exchange the position of the two, helping to balance the population of both.

Slowly lift the frame straight up, so you don't trap, roll, and kill bees between the comb surfaces. If you think you're slow enough, think again. This is the hardest part of examining a colony: care and speed. Too fast and bees get killed.

Pry up the inner cover with your hive tool, opening it a couple inches, and give another half puff or so of smoke. If bees are flying out, give a couple puffs. Slowly remove the inner cover and set it on the cover. Now, before you go any further, do this trick. With a magic marker draw a line on the end of all the top bar lugs on one side of the box. That way you'll always know how to put the frames back in when you remove them. Look down between the frames. What do you see? Between the middle three or four frames will probably be lots of bees and some comb built up. Perhaps nothing has been done in this box yet, if the package is still young. If there is nothing, lift up a corner of the box and puff two or three times, then slowly remove that box and put it on the inner cover.

There should be lots of bees and built-up comb on the center frames in the bottom box. If you are going to look for eggs or perform any other broodnest inspections, follow these points:

- Stand behind or to one side of the colony.
- Using your hive tool, loosen the frame closest to the edge of the box, or the one second closest. If there's a difference, choose the side that has frames with no or very little comb built on them. Loosen both ends of the frame if they are stuck.
- Puff some smoke if bees are coming up between frames and crawling on the top bars and your hands. Watch for lots of bees lining up at the top of the frames, watching you.
- When both ends of the frame are free, lift one end with your fingers and the other with the corner of your hive tool if there is lots of comb, or with your other hand if the frames are not stuck.
- Keep your hive tool flat in the palm of your hand and between your thumb and forefinger, held in place with your little, ring, and middle fingers. This grasp leaves your thumb and forefinger of each hand to hang onto the frame when you lift it up slowly.
- If the frame is empty of comb, carefully set it down on end at the front door so any bees can easily find their way home.
- Loosen the next frame if it's stuck by inserting the curved end of your hive tool between the two frames and twisting it. The leverage this position gives you is amazing, and it will loosen almost any frame. A Maxant-style tool with the hook on one end works well here, too, by sliding the curved end under the top bar, hooking the notch on the frame next to it, and using the whole tool as a lever. You can move mountains with either tool.
- When the frame is loose, remove it slowly, especially if there's comb on even one side of the frame. Look between frames to make sure there isn't burr comb sticking out that will get caught and squeeze bees. Lift the frame straight up until the bottom is clear. This reduces the chance of rolling

When reassembling your hive, be sure the frames are straight up and down so you don't crush bees when the next frame is replaced. Use your hive tool as a lever, and the adjacent frame as the fulcrum to straighten the bottom first, and then the top.

To make sure your frames go back facing the correct way, and in the right location, simply put a line on the top of each, so they line up when assembled.

🐝 *When bees begin looking up at you from between frames, gently puff one or two small puffs of smoke so they go down and leave you to your tasks.*

🐝 *Let the sun shine from behind you, directly over your shoulder, so that it shines down into the bottom of the cells.*

bees (or the queen) between moving combs and crushing them. If there is an obstruction, come at the frame from the other side so that you move the frame that would have been gouged by the burr comb out of the way.

- Hold the frame by the lugs or shoulders between your thumb and forefinger.
- If there's no comb, lean the frame against the first frame, by the front door.
- If there is comb, you'll want to see if eggs or brood are present. Turn so the sunlight is coming over your shoulder. Hold the frame at midchest height and away from your body a bit. Tilt it so the light shines directly down to the bottoms of the cells. Sunlight is the best light there is for seeing eggs or brood, but sometimes one of those little lamps you wear on your forehead can be very helpful, especially for old eyes like mine. If at all possible, hold the frames over the open box or the first box you removed so that if by chance the queen is there, she won't be lost if she falls off the frame.
- If you are going to look at additional frames, slide this one over to fill the empty space left by the two empty frames, keeping everybody inside and safe.
- Puff more smoke if bees are rising between frames or are starting to fly.
- After examining the next frame as before, slide it over out of the way and examine the next frame.
- When the examination is complete, carefully and slowly slide the frames back to their original position. Don't rush, and avoid squishing bees. When the frame is in place, look down to make sure the frame isn't slanted, with the bottom sticking out. If it is, use your hive tool braced on the top of the adjacent frame to align the bottom with the top so the space for the next frame is equal top to bottom.
- When everything's back in place, quickly puff smoke on the edges to remove those bees.
- Slide, rather than drop, the box back in place, pushing bees out of the way. You will probably squash a bee or two, but if you're slow and careful, you won't catch many.
- Replace the inner cover, feeder, empty box, and cover, and call it a day.
- As interesting and exciting as it is to watch bees go about their business when you're examining a colony, keep your visits to ten minutes or less depending on the day, of course. Cool, cloudy days require short work. Warm, sunny days can stand a bit more time. At any rate, after a while, even the most tolerant colony begins to take a dim view of all this exposure and becomes less easy to work.

Honeycomb and Brood Combs

The frames you use in your boxes in which the bees have their broodnest are used much differently than the frames in which they store honey. Three days after an egg is laid, the shell dissolves and a larva emerges. Immediately, nurse bees begin to feed her and the cell is filled with worker jelly for her to eat. Shortly before she pupates and spins her cocoon, she voids her digestive system into the bottom of the cell. When she emerges twelve days later as a fully formed adult, she leaves behind the pile of waste (called *frass*) along with the remains of her cocoon. As soon as she leaves, house bees clean out as much of this as they can, but they cannot remove it all. What remains is sealed with a thin layer of propolis and wax. As a result, after only two or three generations, these cells begin to darken. After a few seasons, they will be nearly black. Added to this is all of the pollen, dirt, plant resins, and other material foragers bring back with them and spread over the dance floor area. Plus, spores from nosema, American foulbrood, and chalkbrood diseases are present in the broodnest, even if you are treating for these maladies. And any pesticides that may have made their way back to the hive are present on or absorbed in the wax.

All of this debris combines to produce a very dark comb, laden with things your brand-new bees don't need to be exposed to and that surely add a level of stress to your colony. Replace older, dark combs routinely to keep the nursery area as clean as possible and to avoid this stress. Every three years, in the spring when most combs are empty, is a good recommendation, but it certainly should occur whenever the comb becomes so dark that when held up to the sun, no light passes through. Two questions should immediately come to mind: How do I know how old the comb is, and how can I check for darkness if I used plastic foundation and the sun won't come though? When you are lining up your frames with that magic marker, write the date on the frame. It's that simple. Three years later, replace it. And removing wax from plastic foundation is a simple task that should be done on the same schedule. When a comb is empty of brood, honey, and pollen (or mostly so), use your hive tool or other scraper to remove the wax from each side. Position the frame so the wax falls into a pail or container, then melt it down, in a wax melter or in a pan on a hot plate, for later use.

Mingling frames from the broodnest with frames from honey supers also causes problems. The material in the cells and the bees walking on this darkened wax will darken the honey stored in the combs and add bits and flavors of what was there before, reducing the pristine quality of the honey you want to harvest. The bottom line: Don't mix frames used for honey with frames used for brood.

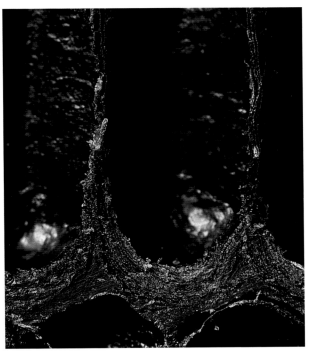

A cross section through an old comb. Note the layers of cocoon, propolis, and gunk. This comb should be replaced.

If your colony is progressing, the queen is working in all three boxes, and the workers are storing pollen and honey around the edges, it's time to add additional room—but not before at least six frames in the bottom two boxes have comb on *both* sides, and the same on at least four frames in the top. Don't rush the colony just because it *seems* to be doing well.

If spring has been late, or if good weather has come in spits and spurts, your colony won't build as fast as an established colony in the same place during the same time. Keep feeding your colony as long as they are taking any of the sugar syrup. They may take it for a few days when it's cool, and then stop when the weather warms, plants are blooming, and the bees can fly.

Sugar syrup may develop a black mold in the pail when the bees don't eat it for a several days, or it may actually ferment if the weather is warm. A good rule of thumb is, if the look or the smell of the syrup is such that you wouldn't drink it, don't give it to your bees. Sugar syrup is an inexpensive and easy way to ensure your bees don't run into the stress of a food shortage, even for a single day.

You don't want the queen wandering up into these brand-new, just-added honey supers to lay eggs and thus darken the honeycombs, so you need to provide a barrier. There are lots of management tricks you can try, but by far the easiest is to place a queen excluder between the top brood box and

The best reason I can think for having two hive tools is that you can clean a messy tool of chunks of propolis and wax easily. Get the big stuff off first, then finish with soap and water using a metal scrubby.

the honey super(s) above it. When the top brood super starts filling, move a frame (or two or three) with some honey (and no brood) up into the super you are going to use for honey. What you are doing is laying some ground rules for management. Above the queen excluder—only honey. Below—honey, pollen, and brood.

The frames with even a little honey above the queen excluder send a message to the food-storing bees that this is an acceptable place for nectar and honey to be stored. Given all this room, and the open invitation to store the food, the food storers will (almost always) begin moving honey up, leaving the three broodnest boxes for mostly brood, pollen, and a little honey.

What usually happens—note I said usually—is that the bees will fill much of the three boxes with brood, a circle of pollen, and honey on the corners and top of the frames in the top box. This honey "cap" usually tells the bees that it should remain brood below, honey above this cap. And they almost always do. It's not a crime to get a bit of brood in your honey supers—as the saying goes, be gladder that the queen is laying than madder that some brood is the honey.

In the warmer regions, this activity is going strong by early spring (by early summer in the coldest areas), so be prepared ahead of time with enough of the right equipment.

Once you've decided that your colony is growing at an acceptable rate, the queen is doing well, food is coming in at a pretty steady rate, the weather has calmed down, and available room for incoming nectar storage is running low (all frames have some comb and none are empty), you can consider adding additional supers for surplus honey storage.

At this point, the sugar syrup feeder has served its purpose and can be removed. If you don't remove it, some syrup may be taken by the bees and stored in the honey super. This isn't a critical mistake, but some of the honey you harvest will be sugar syrup, not floral nectar.

You *can't* have this happen if you are selling honey, but the bees don't care one bit. Remove the feeder, clean it well, and store it for future use. No matter where you are during the nectar flow, be mindful of the wrinkles that can occur. Check the broodnest every week or ten days, and the honey supers at the same time. The greatest stress on your brand-new colony the first season isn't going to be the common diseases and pests you will eventually encounter but, rather, the immediate challenge of establishment. Your goal the first year isn't a crop of honey, but a healthy, well-established colony. Next year, and for years after, your goal is management for production.

This doesn't mean, however, that you should ignore those problems that can occur, and now is a good time to learn the signs and symptoms because the colony is small and easily manipulated.

Integrated Pest Management

In the first edition of this book I examined the problems that can befall an individual honey bee or an entire honey bee colony and suggested a range of techniques to deal with each of them. The least offensive of these techniques and safest for the bees of course were those that prevented problems from occurring in the first place. The best way to avoid a lot of problems our bees have is to use honey bees that are resistant to, or tolerant of the pests they encounter. There are honey bee strains that practice hygienic behaviors, some that are vigorous groomers that remove pests, others that seek out infested larvae even when capped, tear them out and discard them. Some have shorter or longer brood times that foil the pests while others are more agile than the pests they encounter and thus can overcome them.

The next level of dealing with honey bee pests and diseases is to use mechanical methods to keep pests from our beehives. These include: keeping pests out of the hives in the first place; trapping them once they enter before their populations build up in a colony; or providing opportunities for them to leave and not return. Resistant bees, traps, and exclusion are techniques safe to use for the bees, the honey crop, and the beekeeper.

The next level of defense is more intense: chemical alternatives, and within that, soft and hard chemicals.

Each of these techniques has its own advantages and disadvantages for each of the pests examined. Not all work for all pests. Moreover, some years these techniques work fine because pest pressure is light while other years different approaches are required because pest pressure is extreme. The variables of biological systems are unpredictable.

IPM Fundamentals

To successfully manage honey bees you need to be able to identify those pests and know what options you have to control them. You need to know what the pest population is and what it will likely be in the near and distant future if no management techniques are implemented. Also, you must know at what density will the population require treatment (the treatment threshold) so it doesn't cause harm (the economic injury level). Also, you must know the options you have to reduce their populations within an acceptable time frame. These fundamental steps are no different from the pre-emptive measures taken for managing garden critters: fences for rabbits and groundhogs, fake owls for birds and gophers, resistant plants for some diseases, plant disposal, and so on.

There's a couple of ways to approach controlling pests and diseases for the backyard hobbyist.

You can begin at the beginning of the beekeeping season, progress from start to finish taking each management technique in stride as it comes along. This works well and makes sense until you try and encompass every management style and technique imaginable (from warm to cold, tropical to temperate, forest to field, desert to city). Some things will get dropped.

Approach each issue in each season as a separate event, reviewing in detail the techniques each may require. You end up handling many things in an encyclopedic manner, so you either know what the problem is and want to find a solution, or you have a problem you don't understand and need to first define the problem, and then find the answer.

There is usually a complex of events ongoing in your beehive that will need attention. As a beginner I found that I dealt better with each of these when working with each as an individual issue rather than a complex of trials.

At the same time, you have to consider the level of seriousness each of these can play. Some pests are merely a nuisance to the bees or to you, while others are lethal without intervention. Chemical use is an integral part of IPM, but it is (thankfully) usually the last resort for treatment.

Understanding Bee Families

One aspect of a colony's apparent resistance to a particular disease is the many different "families" that exist in any colony at the same time. Recall the queen mates with many different drones during her several mating flights. Reputable queen producers provide drone source colonies in the area where they keep their queen mating colonies. These drone source colonies are not genetically identical, but rather are diverse in their backgrounds. Each has a particular strength as part of its makeup: for instance, one may contribute a strong hygienic behavior background, another may have that particular color you are interested in, another may be from genes that are particularly good at pollen collection, another at nectar collection, another at defense, another at longevity—the list of positive attributes goes on and on.

Most, perhaps all, of these families are present in a colony at any one time, so a very small portion of the population is doing all or most of the hygienic cleaning of diseased larvae, while another is meeting you at the front door, and other families are so busy collecting nectar that they never get to be guards, house cleaners, or pollen gatherers.

You want a colony that has a large, diverse collection of behaviors ongoing at the same time so the colony as a whole is the best it can be at every aspect of survival. But realize also that some of these behaviors may not be present in a colony all the time. During mating the queen may mate with a single drone that has the pollen collection drive, and with many that have strong hygienic behavior, thus your colony will be safer from disease, but have to work harder at pollen collection.

Maladies

The first season you have bees it is unlikely you will encounter a serious disease problem. Some pests may become a problem later in the season and if you have bought used equipment some of these diseases may show up. But mostly, new bees on new equipment have a honeymoon before the problems of the world come home. By the fall, during the coming winter and by the second spring, however, you need to be most aware of all the bad things in the world. But still, some may show up right away and to be safe you should know what they are and what to do if they come to visit.

Nosema

Nosema is a protozoan-caused disease spread by *Nosema ceranae* spores that attack only adult bees after they have consumed the spores. Once consumed, the spores germinate inside the bee and send a long, coiled stylet through the wall of the bee's midgut and begin consuming the haemolympth of the bees, which is, essentially, the blood of the bee. This nourishes the *Nosema* protozoa so that it matures and produces more spores. Some of these spores remain and continue to infest the host bees, while others are expelled in fecal material, thus spreading the disease.

Symptoms are not easily observed, especially in the early stages, and it can be a difficult disease to identify. The most certain way to know if your bees have the disease is to send a sample of bees to the USDA Honey Bee Research Lab in Beltsville, Maryland (see Resources for instructions). They will examine your bees and tell you a lot about them—the presence of *Varroa* mites, the number of *Nosema* spores, and the presence of American foulbrood disease, and whether the disease is resistant to the commonly used antibiotic. Because of the seriousness of nosema, here's a recommendation you probably won't hear anywhere else: Send in a sample of bees right after you get your package. The lab will tell you the average spore count per bee, which can range anywhere from 0 to several million.

The best approach is to assume your bees have a mild infestation and to combat it the best way you can. Since the protozoa attack the midgut of the honey bee she tends to stop eating and eventually starves to death. Before the bee dies, the protozoa consumes the interior of her digestive system, matures, reproduces and spreads more spores.

The best way to combat this disease is to make certain your bees continue to eat a healthy diet. To do this, start by making available lots of honey and fresh pollen or pollen supplement to a colony that seems to be listless and not eating and especially to a colony that isn't foraging well when others are.

If that in itself doesn't inspire them, try a feeding stimulant that makes their food more attractive!

There are several feeding stimulants on the market that all contain various essential oils and herb extracts. Lemongrass oil, thymol, and other essential oils all have basically the same effect—they stimulate even infected honey bees to eat, and eating combats the disease. Continuous feeding at the beginning of a season has been shown to reduce spore counts, and even reverse the disease. Even bees not infected will eat more and remain healthier because of this.

The antibiotic fumigillin, used to stop spores from germinating, was prescribed for nosema, but it has been found that although it stopped spores for a time, once the antibiotic was no longer in the bee's system, the protozoa actually reproduced even faster, causing rapid and lethal colony decline. Beekeepers have found that the essential oil feeding stimulants actually have a much better result, and this is currently the only way to combat this disease.

Another recent discovery is that a colony infected with nosema that is handling the disease by being fed good food and essential oil additives in its food sees a huge and rapid increase in spore production when exposed to even tiny amounts of agricultural insecticide. The pesticide seems to weaken an already stressed organism past its ability to continue handling the disease.

One thing to keep in mind: If your queen encounters nosema, she's a goner. The disease is especially hard on queens because it wreaks havoc on their ability to lay eggs. If she has the disease when she arrives or gets it soon after from being released in old, spore-infected equipment, she will be gone in just about three weeks and your colony will be queenless, unless the first queen was first able to produce fertilized eggs and the bees are able to produce queens from those. If that happens you will see queen cells almost immediately in your brand new package colony. Plan ahead and get a queen ordered.

Tip: Feeding Stimulants

It's a good idea to use feeding stimulants in your sugar syrup when feeding packages right away. Nosema and some of the other maladies are encouraged to attack when bees are under stress: cool, rainy spring weather; moving; accepting a new queen; and inadequate or uneven food sources lend themselves to stress and *Nosema* infestation.

Emergency supercedure cells: Your bees build them when they have lost their queen suddenly (such as in a case of nosema). The bees select a larva that is less than three days old, and begin to produce a queen cell that hangs down from the cell the larva was in, hanging between the frames, to accommodate the larger queen larva that will eventually form in this cell. These may occur for any of several reasons, and they are always a warning sign.

Avoiding this disease is the best choice. Ask the package supplier what they are doing to control this disease in their operation. Ask if they treat and if they feed these essential oil products to their queen mating colonies to protect the queens.

Once your bees arrive, make certain they are given lots of food and to ensure they eat, add a feeding stimulant at full strength.

Otherwise, get your queen released, don't open the colony more than you have to, and add protein to their diet in the form of a supplemental patty. Take these further precautions: Reduce drifting so this or other diseases don't spread between your colonies, make sure the colony is getting as much sun as possible, and make sure the colony has access to water.

These activities should reduce or eliminate the need to use medication for nosema.

It is recommended that you test your bees for this infestation by sending them to the Beltsville Bee Lab, who will give you a spore count per bee report. At this time, good recommendations for treatment are a mixed bag. Commercial beekeepers are feeding their bees with a concentrated sugar solution containing one of the essential oil products. The oils have some effect on the *Nosema* organisms, plus they encourage the bees to eat more.

Diagnosis without a spore count is difficult as there are no easily visible signs of the infestation occurring. A colony slow to build when others are thriving is a sign, as is a colony with lots of brood but not enough adults to deal with the brood.

There is a field test for spores, requiring a microscope, and it will pay you to eventually learn this technique. Your local club may own a scope, or you may have to purchase one. But for now, have your bees tested, feed some of the essential oil to your bees in the spring, during any dearth in the summer, and before winter sets in to get good food, and oil, into your bees.

When you send bee samples to the Beltsville Bee lab, the scientists there will remove your bees from the bag containing bees and alcohol and grind them up using a mortar and pestle. The solids are strained out and the liquid placed under a microscope. The sample is in a grid and counting spores in a single grid provides a good estimate of how many spores per bee were in the sample. A heavily infected bee will have between 7 and 10 million spores.

You'll first see chalkbrood mummies on the bottom board, having been removed by house bees.

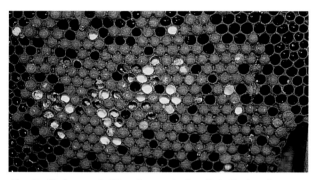

You'll also see chalkbrood mummies in cells in the broodnest. They will be white, chalky, and hard. They may rattle when you move the frame. Some may have black spots on them, meaning the disease has matured to the point of producing spores, providing additional means of infestation.

Chalkbrood

Chalkbrood, *Ascosphaera apis*, (usually called chalk), is a fungus that attacks the larval stage of the honey bee. It is a disease that shows up when the colony is stressed, especially with food shortages and erratic temperatures. Damp, cold conditions don't help the bees any, and keeping a colony in a high, dry location and providing as much ventilation as possible is often recommended as a way to reduce incidence of this disease. Spring is the most common time of the year for this to show, but it can, and will appear at any time. Recall the genetic diversity of the worker population that may keep this disease completely in check or allow it to run rampant any time of the year.

Somewhat like nosema, chalk is spread by spores from previous infections. Spores are first encountered when bees from your colony rob an infected colony, bees with spores drift into your colony, or you exchange equipment from an infected colony with one that isn't.

Adult bees that have cleaned up larvae that have recently died of chalk inadvertently feed spores to other larvae. The spores are ingested, then germinate in the gut and compete with the larva for the food. Very soon the fungus will cause the larva to starve to death. It then invades the tissue of the larva, consuming it and spreading its hyphae throughout the dead larva's body tissue, consuming it all. Eventually the body tissue turns hard and is chalky white. This is called a *mummy*. If conditions are favorable and more than one strain of the fungus is present, when mature, the fungus will mate and reproduce, casting off black spores; the dead larva now becomes white with black splotches on it. When nurse bees go to remove the chalky dead larvae, now hard and rattling around in the cell, they get some of these spores on their bodies and then share them with young larvae, completing the circle. It is a serious problem when a high percentage of the larvae in a colony die because of it.

Avoiding Chalkbrood

- Strong colonies in the spring maintain even internal broodnest temperatures and can collect adequate food.
- Replace brood combs every three years or so to keep the number of spores to the very minimum.
- Remove and destroy entire brood combs that have lots of infected larva.
- Maintain bees that have hygienic behavior.
- As a matter of course, keep stress to a minimum with lots of available food, a healthy population, and requeening when necessary.

The best defense is to have bees that are extremely hygienic who will clean out dead larvae before the fungus matures and produces spores. This pretty much stops the infestation. For a brand new package on foundation the likelihood of this developing the first season is slim, but possible. Later in the summer and certainly next spring be on the watch for it though, and if feeding and time don't reduce or eliminate the problem consider requeening your colony with a strain of bees claiming to be more hygienic than the norm. There are no chemical cures or treatments for this disease. Good management and resistant bees are the things that work.

European Foulbrood

Yet another stress disease, the bacteria *Melissococcus pluton* is associated with springtime stress. Much like chalk, spores are fed to young larvae by nurse bees. Spores germinate in the midgut of the larva, aggressively and rapidly compete for food, and the larva starves at a very young age. The bacteria consume the larval tissue until all that's left is a twisted, brown rubbery mess, usually still curled at the bottom of the cell. Sometimes they are stretched out lengthwise along the side. These "scales" are removed by nurse bees who then come in contact with the spores.

Like chalk, spores are first encountered when bees from your colony rob an infected colony, bees with spores drift into your colony, or you exchange equipment from an infected colony.

Your first indication of an infestation is a brood pattern that appears spotty. Careful examination of an early infestation shows dead and dying larvae in the very bottom of brood cells that ARE NOT SNOW WHITE! Repeat. The fastest sign of a problem with larvae is that brood cells are no longer snow white. A darker color is always a bad thing and needs to be attended to. After the larvae die, they appear as scales— lying along the length of the cell. You can easily remove these with a toothpick to examine further, but the key is recognizing the color change and quick removal.

Consider treatment the same as for chalk. Use hygienic bees, reduce stress to a minimum, make sure adequate food is available, provide good ventilation, and make sure that the colony gets as much sun as possible. Generally a good honey flow, warm weather, and a healthy population of hygienic nurse bees cleans up European foulbrood (EFB). You most likely won't see this the first season you have bees from a package, but, like chalk, watch for it the next spring. Requeening is always a good idea if this becomes a persistent problem.

European foulbrood–infected larvae are first noted because they are tan to yellowish, turning dark brown and eventually to black. The larvae die before the cell is capped, which is one of the tell-tale signs used to distinguish between this and American foulbrood, another bacterial disease. Plus, the remains are easily removed by the bees, thus somewhat reducing the incidence of further infestation.

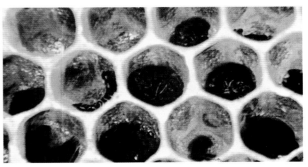

European foulbrood–infected larvae are dark brown to black, rubbery and sunken to the bottom or along the side of the cell. The larvae die before the cell is capped, unlike American foulbrood-infected larvae, which perish after the cell is capped.

American Foulbrood

American foulbrood (AFB) is caused by the bacteria *Paenibacillus larvae*. This is by far the most destructive disease honey bees get. AFB is spread by spores that are consumed by larvae during their first couple of days in the cell. After that, they become immune to the disease.

You will probably never encounter this disease in your colonies. It is rare and well policed, but it is so very serious we need to spend some time with it, first, so that you can recognize it if it does show up, and second, so that you can efficiently deal with it so that it doesn't spread.

Spores enter colonies in a variety of ways:

- Workers rob a wild colony, another apiary, or another colony in your apiary, and bring home spores inadvertently through contact or in the honey they stole.

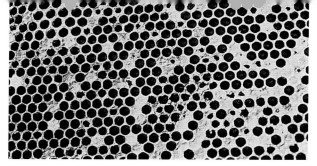

🐝 *A spotty brood pattern is always a sign of trouble, and is a first, best sign of American foulbrood disease. This is your most obvious clue that a disease is present.*

🐝 *One diagnostic technique for American foulbrood is what is commonly called the rope test. Shortly after the cell is capped the larva dies and its body becomes jellylike. Poke a small twig or toothpick through the wax capping and carefully stir it around. Withdraw the probe very slowly and the sticky mass of dead larva will cling to the probe. The color will be coffee brown to dark gray, and it will draw out of the cell ½" to 1" (1.3 to 2.5 cm) before it snaps back. Though distinctive, this isn't the definitive test for AFB. If the rope test is positive, you should check for signs of this disease in all your colonies.*

- Equipment previously contaminated is sold and reused.
- Frames are moved from a colony with AFB to one without it.
- Gloves and hive tools used when working a colony with AFB are not cleaned before use in another colony.

Spores are fed to larvae fifty or so hours old or younger—a very small window—and the infection begins. Older larvae have a built-in resistance and are not affected. Once ingested the spores germinate and begin consuming the larvae, much like the other diseases mentioned. The larvae eventually die *after* the cell is capped. Before they die, the larvae turn a brownish/yellow color. This is a sure sign that something is wrong in a colony. Larvae should always be a stunning, nearly neon white. And they should be shiny and glistening, not dull.

Once the cell is capped the infected larvae die and turn into a jellylike mass. Shortly after the cell is capped the larvae, at first standing in the cell preparing to pupate, are now spread along the side of the cell, drying down to a hard, nearly black scale that is difficult to remove even for the beekeeper, let alone a house bee trying to clean out the cell. The dried scale produces millions of spores that are picked up by the cleaning bees and inadvertently fed to additional larvae. Thus, the disease spreads within a colony. This disease will kill a colony within a season.

Because of the seriousness of AFB, most departments of agriculture have regulations on containment, treatments, and controls. There are a couple of antibiotics (a tylosin product is one) labeled for use when a colony is officially diagnosed with this disease (from samples taken by a bee inspector and confirmed by a recognized diagnostic lab, such as the USDA). You cannot use either to prevent the disease, just as you would not take an antibiotic to prevent a disease you did not have. Resistance to antibiotics is a worldwide problem and this is one way to prevent that from occurring here. With no medication to control this disease, and since even good hygienic bees and stress reduction does not help, what's the answer to AFB? Extreme beekeeper hygiene. Careful examination and diagnosis can locate symptoms early so that brood frames, and entire supers if necessary, can be destroyed. Frames with diseased larvae can be removed and burned, but recall that the larvae don't die until the cells are capped, so detection isn't easy, especially when the infestation is small and early.

You can assume that in even a mildly infested colony, most of the broodnest-area bees—those that feed and clean—have AFB spores associated with them. Further, these are some of the bees that take nectar from foragers, passing spores to them also.

Avoiding AFB

American foulbrood annually infests between 2 and 5 percent of all colonies in the United States. Considering the ease with which it is spread this very low rate can be attributed primarily to early detection and destruction of infested equipment by both beekeepers and state department of agriculture–sponsored inspection services.

Here are some tips for avoiding AFB:

- Never, ever buy used equipment—no matter how well you know the beekeeper, no matter how attractive the price, even if it's from your brother.

- Make sure your colonies are registered with your state inspection service (see Apiary Inspectors of America webpage, www.apiaryinspectors.org) so they are inspected by a trained inspector who can help with identifying problems and suggest treatment options legal in your state.

- Seek out queen producers who sell queens that generate bees with strong hygienic behavior.

- Routinely inspect the broodnest (every ten days to two weeks, minimum) looking for spotty brood patterns, brood that's not glistening white, and sunken cappings.

- If available, contact your local inspector if you find suspect cells (see photos for examples).

- In the unlikely event that you find a suspect infestation, immediately remove that frame from the colony and isolate and store it so other bees don't have access. Freezing it is the best choice, but double wrapping in a plastic bag is acceptable until you can positively confirm the disease is present or the frame is clean.

- If inspection services aren't available, find a local beekeeper who is skilled in identification for an opinion, and then submit a sample to the USDA (if you live in the United States), or another government agency. (See see Apiary Inspectors of America webpage, www.apiaryinspectors.org).

- If AFB is confirmed your only good option is to burn all of the frames, inner cover, and bottom board. Scrape the insides of the boxes, especially the frame rests, completely clean of adhering wax and propolis and then scorch the insides and frame rests with a propane torch until the wood just starts to turn brown. Any less and the remaining spores will not be destroyed.

If you choose to treat an infected colony with an antibiotic, you open the door for a permanent source of infection because the medication does not destroy spores. Thus, you need to treat every spring and fall with an antibiotic to keep the disease from reappearing. It is an ongoing, losing proposition.

If you choose to burn infested equipment, dig a hole large enough to accommodate much of the equipment. The hole will contain the fire and hold the ashes and any honey that runs from the burned frames. Then, kill the bees by dumping a 5-gallon (19 L) pail of soapy water directly into the colony from the top. Use 5 cups (1.2 L) of dish soap in the pail of water. When the bees are dead, first put in frames and burn them, dousing the equipment and starting the fire with a petroleum-based charcoal lighter, then add the inner cover and bottom board. When all the equipment has been consumed, cover the ashes completely so other bees won't find the equipment and honey and become contaminated. This is a horrendous loss of equipment, so avoiding AFB is certainly recommended. Before burning, check your local fire safety laws. A burning barrel also works for this and is much, much easier to organize and manage. But no matter how you do it, burning is necessary.

Healthy larvae and pupa, such as those shown here, are pure, glistening white.

For American foulbrood infestations look for scales, the dried remains of the dead larvae in the cells. They are hard and difficult to remove. Here, they are shiny black, but sometimes they are dull black. Usually the bees can't remove them, but they try, and in the process pick up spores, spreading them around the colony.

The Value of Early Detection

If you missed early detection, you can be fairly certain that most brood frames have several to many infected brood, that most of the nurse bees have been contaminated, and that some percentage of the field force is contaminated also. This means when you destroy a colony's equipment you must destroy the bees within.

Detecting early signs of AFB, or any brood disease for that matter, begins with looking at the brood pattern and at the cappings themselves. A spotty pattern could be an indication that something is amiss. Maybe it's only a young queen just getting started or many workers emerging very close to each other. But look closer at the cappings if the pattern is spotty. Are they just slightly rounded, convex, with the curve upward? Are they all the same color and texture? Do some have small holes in the center and are these sagging downward, sunken, rather than upward? The capping in an advanced AFB infestation can have a greasy appearance rather than the normal rough, waxy appearance. Any of these symptoms tell you to look further. Using your hive tool or a toothpick-size twig, uncap several cells with symptoms and several without to see if there is a difference. Diseased brood—whether by AFB, EFB, chalk, or other diseases—are *not* white. The colors may range from translucent to tan to brown to nearly black. Your first clue, and the key to early detection, is the presence of larvae in their cells that *are not* pure, bright white.

Once an infected larva's cell is capped and the larva dies, the wax capping over the cell changes. Rather than remaining just slightly convex, it sinks in and will usually change color.

Some house bees with hygienic behavior traits (recall the many families in a colony—some have hygienic behavior, some don't) may begin to investigate this odd cell by opening it up at the center. If present, other bees with strong hygienic behavior will then remove the diseased larva before the bacteria have a chance to form spores.

This is where an effective integrated pest management program gets tricky. Because of regulatory zero tolerance for AFB, an infestation must be treated or destroyed. It's my opinion that the antibiotic route is unacceptable because once applied it is a life sentence for the equipment, the beekeeper, and the bees that live there because the equipment is contaminated forever. However, a strong hygienic colony can tolerate a light infestation of AFB by keeping it cleaned out. Destroying a colony like this removes those hygienic genes from the population and leaves only bees that are susceptible to AFB.

Tracheal Mites

Tracheal mites, *Acarapis woodi*, are microscopic in size and cannot be directly observed. These mites are considered to be universal, though in much of the world susceptible strains of bees have been eliminated, and mostly, repeat, mostly resistant or tolerant strains are all that are left. Just enough susceptible bees remain, however, that you can't completely eliminate them as a cause of a colony's demise.

Their cycle is typical for parasitic mites. Fertile female mites enter the main thoracic trachea (breathing tube) beneath the wings of young bees—queens, drones, and workers are all susceptible—lay eggs inside and produce young. Inside the tracheal tube (about the thickness of a human hair), the first egg is a male, and those following are female. The male mates with his sisters while still in the tube. The emerged mites pierce the tube and feed on the bee's blood, the haemolymph, that bleeds through the wound.

If infestations become excessive an infested bee has a much shortened life, colony honey production suffers, and during the winter, most or all of the bees will die. Perhaps, in the

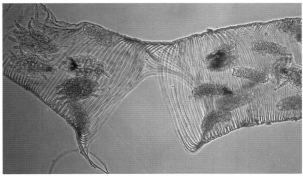

Tracheal mites live inside a honey bee's trachea. A trachea, or breathing tube, is roughly the diameter of a human hair.

Put the grease patty directly on the top bars, off center, and let the bees consume it. While doing so, they'll pick up minute amounts of the grease, foiling further mite infestations.

spring, you will notice bees crawling near the entrance, but perhaps not.

Tracheal mites have been present in the United States for nearly thirty years, and nearly all of the strains of bees susceptible to these mites in the United States are gone. Queen producers did little to select for resistance during the first decade they were here and bees died by the billions. At almost the same time *Varroa* mites came to the United States; the two combined have been disastrous.

No matter where you live, an effective treatment for tracheal mites, called a grease patty, is made as follows:

- Obtain a can of solid vegetable shortening—3 pounds (1.5 kg) is common.
- Get a pan large enough to easily hold twice this amount of shortening.
- Put the shortening in the pan, and slowly warm it on your stove.
- When the shortening reaches the translucent stage, and is not quite liquid, begin adding 10 pounds (4.5 kg) of regular sugar. (The formula is 3:1, sugar to shortening.)
- When all the sugar is added, add an additional half pound (0.2 kg) of honey (from your bees or from a safe source). Stir in.
- Turn off the heat, and add 1 ounce (28 g) of food-grade peppermint flavoring. Stir and mix well.
- Let cool to room temperature.
- Using an ice cream scoop and waxed paper, scoop out about a hamburger-size dose of the finished mix and put it on a sheet of the waxed paper about twice the dose's size. Add more paper and flatten. Freeze the patties until needed.
- Place one patty on your colony (leave the paper on the bottom) between the two boxes with the most bees in them, as early in the spring as you can, and replace it when most of it is gone. Continue adding these until your honey flow starts,

then quit until the honey flow is over.
- Add a patty in the fall, and have one in place for overwintering.
- This recipe makes a lot of patties, so you may end up sharing with a friend.

Because sugar and shortening are essentially odorless to a honey bee, the honey and peppermint act as attractants. You will find some colonies eat these patties rapidly, whereas others are slow. If the bees are slow, add additional honey, sugar, and peppermint to a new mix to increase attractiveness. A rare few will never eat them. If that happens, leave them on anyway and hope for the best. It's all you can do for these bees.

Grease patties work well because the shortening, when on a young honey bee, confuses newly emerged female mites looking for a new host, and the rate of infestation in young bees drops precipitously.

Occasionally a colony will succumb to these mites, however, especially if you are not using volatile *Varroa* mite treatments. This occurs most often during the winter and very early spring. You find in your colony lots of honey and few or no bees. Sometimes, you will find them in two, three, or more small clusters in different parts of the hive.

Remove the colony from outside after brushing away any remaining dead bees until you are ready to add new bees. But be certain the colony didn't perish from a disease (AFB) before using the equipment. There are no remaining mites in a colony such as this so recontamination on that count is not a problem.

The grease patty is good insurance in the fall, when it is placed on the colony after most brood rearing has ceased. It should be left on until the spring. There is little new good data about this since no one is looking at this problem any longer. By using these patties you are again selecting bees that are still susceptible to these mites and protecting them. Ask what others in your area are doing.

The Russian lines and survivor bees are very tolerant to tracheal mites and they should not have a problem with them at all. An uncommon strain of bee called *Buckfast* are highly resistant also.

Resistant bees; good management practices to reduce stress, diseases, and other pests; and perhaps using grease patties in the fall should keep tracheal mite populations in check.

Varroa Mites

For nearly thirty years *Varroa* mites and the problems that arise when they are present have been the most challenging health problem honey bees and beekeepers have had to deal with. And during those thirty years, things have gotten worse, and better, relative to the damage the mites cause and how we deal with them.

For this reason the following section is detailed and perhaps more lengthy than others in this book. But the importance of handling *Varroa* mites in your colonies cannot be stressed enough. Controlling *Varroa* mites in beekeeping is as important as knowing which side of the road to stay on when learning to drive. Mostly, if you don't control *Varroa*, your bees die. But there are ways to avoid *Varroa*, resist *Varroa*, or treat *Varroa* that are beginning to make headway. But first, you should know what *Varroa* is.

Like tracheal mites, this pest is nearly universal, and like AFB, it is usually deadly without beekeeper intervention at some level. But as with both of these pests, eliminating a colony or chemically treating the problem takes the issue of maintaining a resistant stock out of the conversation. We'll deal with this issue later, however. Fortunately, unlike tracheal mites this pest is large enough to see, so detection isn't guesswork.

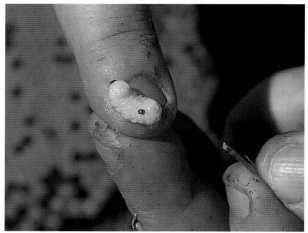

🐝 *If you remove a larva from its cell that has a* Varroa *mite already in it, this is what you will see. Note the size and color of the* Varroa *mite. They are easy to identify. Generally, when you remove the pupa from the cell, the mite will scurry away, and they can move fast.*

In fact, there are two stages in a female mite's life. One, called the phoretic stage, is when a mated female, released from the confines of the cell she was born and mated in, finds an adult bee to live on. The flattened shape of these mites is ideal for slipping underneath the gaps between a bee's segments, almost invisible to the beekeeper and almost impossible for the bee to remove. They can pass from bee to bee within a colony or ride to other colonies through drifting or robbing. With brood present, this period lasts from four or five days up to a couple of weeks, but if there's no brood, they will ride bees for as long as half a year. During this time they puncture the soft intersegment tissue and feed directly on the haemolymph of the adult bee, compromising her immune system and sharing as many viruses as are present in the colony with all the bees in the colony. They are most exposed during this time and chemical treatments are most effective. Normally, a female mite lives about four weeks when brood is present, and six or more months when not.

During the next phase, the reproductive portion of her life, our fertile female *Varroa* mite seeks out and finds a cell with a larva that is just ready to be covered, or capped, by the nurse bees. She scoots down to the bottom of the cell and hides there, in the remaining food and frass, until the cell is capped.

When the cell of a drone or worker is capped, the mite emerges from the puddle of food below the larva and is ready to begin feeding. She climbs on the larva, finds just the right spot, and uses her extremely strong piercing mouthparts to puncture the cuticle so she can feed on the haemolymph, or blood, of the larva. In doing so she passes into the wound salival compounds that compromise and weaken the immune system of the bee, so instantly the bee becomes less able to fend off attacks by other diseases, viruses, and agricultural chemicals. Her damaged immune system that until now had been handling, mostly, the many viruses that are latent in a beehive is no longer able to do so, and any of several of these killing viruses can emerge, including Kashmir, Sacbrood, Acute paralysis, Israeli paralysis, and Deformed wing (the most noticeable of these).

After three days, the female mite, now energized with a blood meal, lays her first egg, a male, and continues to feed. About thirty hours later, she lays her next egg, a female, as is her next if she has time. A female mite produces on average 1.3 female mites if she is in a cell with a worker, but can produce as many as five females if she is in a cell with a drone. You can see how the mite population on average, can increase twelvefold if the colony has brood for about six months. Without intervention and where there is brood year-round, the population exceeds an eight-hundred-fold increase.

The female and her offspring feed on the now-pupal bee, all at the same initial spot chosen by the female. When mature, the brother mates with his sisters. The mites continue to feed until the now-damaged adult bee emerges from her cell. It takes just under six days for a female mite to reach maturity, and about six and a half days for a male mite to do the same.

The damage done to this bee is incredible. She already has suffered or is unable to resist the attacks of several of the viruses that she is now susceptible to. This damaged bee is less able to do her normal jobs in the colony—house cleaning, feeding, guarding, and foraging. Her life is shortened, and sometimes the damage is so great she doesn't even emerge, but dies in the cell.

If she does emerge and begins life as a house bee one of her first jobs is to feed larvae. Her glands to produce food are less capable of doing so, and when she does feed, she passes the viruses she has on to the next generation as well as the queen when she feeds her. The queen in turn can pass these deadly viruses on in her eggs. As she ages, our wounded worker shares her disability with all manner of individuals in the colony, and soon every bee in the colony has some level of several of the viruses. Once this level is reached the colony is doomed.

But there's more to this story. The most measurable result of this attack is that adult bees do not live as long as their unattacked peers. They die two, sometimes three or four weeks, sooner than an uncompromised bee. This adds up in a colony, and as adult bees that should be attaining the age when they are able to forage are dying, younger and younger bees take up the slack, but they, too, are dying. At some point, more bees are dying than are being born, and the population of the colony begins to rapidly dwindle.

When young, newly mated female mites emerge with the drone or worker on which they had been feeding, and while in the phoretic stage, they seek adults until they find new cells to invade so they can repeat the process. The male mites do not leave, but die in the cell. No matter what treatment you choose, when females are exposed, they are at the weakest, most vulnerable time in their life cycle. You must control *Varroa* mite populations or they can kill the colony. Fortunately, several treatments are available, and monitoring mite populations and treating is a routine part of *Varroa* mite management.

How to Monitor Mite Populations

To win the *Varroa* mite battle you first must know how many mites are in your colony. Knowing that number will tell you if there are so many there that you need to take action, or if your avoidance techniques are paying off and you can relax. Count mites about once a month during the season.

There are a couple of different sampling techniques used to determine the presence and number of *Varroa* mites in a colony. They each begin with a sample of about 300 bees from

your colony. To gather the bees, remove a frame from the brood box. Examine for the queen. If she's there, remove her. Quickly shake the frame of bees into a dishpan. Check another frame and repeat. Three hundred bees amounts to a ½ cup. Once the bees are in the dishpan, quickly bump them to one side. Shake another frame and bump them again. Take a half-cup measure, scoop out to fill the measuring cup, and dump them into a quart jar and cover. Thump them to the bottom and begin your sampling technique.

The first sampling technique option is the ether roll. Lift the cover on your quart jar full of bees a half inch (1.3 cm) on one side and spray with automobile starting fluid (ether) for a full second. This kills the bees. As soon as you close the jar, begin rolling it slowly. Roll for exactly five minutes. Not four, not six. When the bees die they regurgitate the nectar in their honey stomachs. This collects on the inside of the jar as you roll it. It also kills any mites clinging to the bees. Most of these mites are caught in the sticky goop on the side of the jar. After rolling the jar for five minutes, count the mites.

Ether roll procedure. Shake bees off of brood box frames into a plastic dishpan. Make certain the queen is not on the frame. Scoop a level half cup of bees into a quart jar with the cover handy. Thump the jar on the bottom so the bees fall to the bottom, lift the lid a half inch or so and put a 1-second spray of auto starter fluid—ether—into the jar and quickly close the top. Roll the jar gently for 5 minutes. The ether will kill both the bees and the mites. The bees regurgitate nectar from their honey stomach which clings to the side of the jar. Dead mites get stuck in the nectar. When finished, count the mites clinging to the glass, and since a half cup is 300 bees, you can figure a percent infestation—more than 5 percent and action is needed.

🐝 A good comparison of size between a Varroa *mite* and an adult honey bee.

🐝 Close up of a female Varroa *mite on an adult* honey bee.

Another technique is to sugar your bees. When you capture the bees, put them in a quart (946 ml) jar with the two-part cap used for canning. Replace the inside lid with a circle of small mesh wire (but larger than window screen). Pour in ½ cup (60 g) of powdered sugar into the jar through the mesh. Roll the bees in the jar for five minutes. The tiny sugar particles cause the mites to lose their grip and fall off the bees. After rolling, pour the sugar and the mites out of the jar through the screen onto a white surface. The mites are easily visible. Count them. The bees can return to the colony, unharmed.

You can also do this using alcohol. Pour in a ½ cup (120 ml) of rubbing alcohol into your jar of bees and shake it for five minutes. This kills the bees and kills and dislodges the mites. Using the mesh top, pour out the liquid and the mites through a filter—paper towels work fine for this—covering another container for the alcohol, and count the mites.

Divide the number of mites in all of these jars by 300—the number of bees that you tested. This is an estimate of the percentage of adult bees outside of cells that have phoretic mites attached. Two to 5 percent is the maximum you want to find. If there are more, treat your colony. Multiply your percentage by the number of bees in your colony for an estimate of how many of your adult bees have phoretic mites. And keep in mind that this only represents 20 percent of the total mites—the other 80 percent are inside cells.

Sticky Boards

Earlier it was recommended that you purchase a screened bottom board with a rear-access removable tray that would slide in beneath the screen. *Varroa* mites are the reason. You'd think that the bees in your brand new package would be pretty healthy, wouldn't you? Why not check and see? Set up your

screened bottom board and slide a sticky board (available from every bee supplier) beneath the screen. Small particles, including mites, that fall from above the screen—from the bees or the frames above—will pass through the screen but be captured on the sticky board below the screen.

Adult *Varroa* mites aren't always reproducing in cells. Those not in cells often spend time on top bars and on the comb waiting for an adult bee to walk by so they can catch a ride and feed on them, move around the colony, find another bee, or even travel to another colony on a robbing or drifting bee.

During this process they may be brushed off, let go, slip and fall, or die. Some end up falling to the bottom of the colony, through the screen, and are caught on the sticky board below. This happens constantly and scientists have for the most part figured out how many will fall every day in small, medium, and large colonies during the spring, summer, and fall. (See the box on page 122 for these estimates.)

Once you know how many mites have fallen, how big the colony is, and what time of year it is (and to some degree what kind of bees are in that colony), you can estimate if the *Varroa* infestation is too large. Most important, you can then determine what you should be doing to change that population. Your decision to take action should reflect the IPM discussion (see page 108), considering the pest population, the economic threshold, and the injury levels.

Using Sticky Boards

The plastic inserts that come with some screened bottom boards can be made into your own sticky boards. Apply a thick layer of petroleum jelly or vegetable-oil spray to the surface of the board. Insert it in the slot that holds the board. Mites that fall will become stuck in the goo.

🐝 Feeding Varroa-*infested drone brood to chickens is a treat for the* birds, but hard on the Varroa, *and the drones.*

This shows the underside of a screened bottom board. The strings hold the plastic sheet of your sticky board in place.

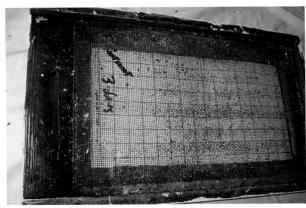

Count the mites immediately if using vegetable oil or petroleum jelly, or cover with clear plastic food wrap if using a commercial sticky medium.

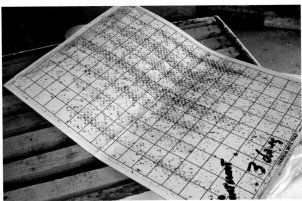

The reverse side of a screened bottom board: Debris and mites fall from the colony above, through the screen, and are caught on the sticky board. When the three days have passed, remove the board.

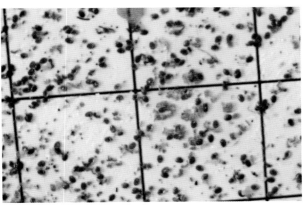

There are hundreds of oval, rust-colored mites on this board. The rest of the debris is wax and other regular colony junk.

To check if the goo is thick enough, put the board in place, then smoke the colony for several minutes. Close the colony, wait fifteen minutes, and pull out the board. The smoke will knock a few mites off. If they are moving, you don't have enough goop. If they aren't, you do.

After it has been on for three days, remove and count the mites on the whole board (don't use your test colony for three days). Record the number in your notebook and remove the goo with a large, flat stick or a spatula. It's messy, but necessary.

Commercial boards have a grid printed for easier counting and they fit exactly in the screened bottom board. Peel off the protective covering, put the sticky board in the tray in your screened bottom board, and wait the three days. During those three days don't visit the colony. Your visit would cause mites to fall that would not normally have fallen. Many are dislodged when you smoke a colony or move frames around. (Try this sometime during the summer to really see how disruptive opening a colony can be.)

After three days, remove the sticky board and immediately cover with clear plastic food wrap. Pull it tight and smooth it down so there are few air bubbles or wrinkles. This keeps the board from accumulating additional material and it allows you to stack them for storage and still be able to count them later.

As soon as you can, count the mites that are stuck on the board. Next, take that total and divide by three. This number represents the average number of mites that fell each day. If you find hundreds you have a serious problem. If you find none you have a great colony. Most likely, the number will be in between, but if you've been reading this book, that number will be much closer to 0 than to 100.

Now look at the chart on page 122 that gives the guidelines of how many mites you find relative to the size of the colony and the time of year. They are based on the number of bees in your colony at that time of year and the size of the colony. The number of mites that fall represents a calculated percentage of all the loose mites in the colony. Knowing that roughly only 20 percent of the mites in a colony are not in cells, a count above the numbers on the chart indicates that the percentage of bees infested in the colony is too high. Those numbers are very conservative, but when the daily mite count reaches those levels you are going to have to take a more aggressive stand on mite control. Study it carefully. If you are serious about your IPM program and have been monitoring the mite levels in your colonies, you will know long before the counts get to these levels that you need additional mite prevention or removal techniques in your world. Know, too, that there are many ways to beat this foe.

Keep using the sticky boards. The number of mites you count on the sticky board (considering time of year and colony size) will tell you if your IPM techniques are working, or if you need to consider using more aggressive methods or finding resistant bees.

Maximum Number of Mite Guidelines

This chart shows how many mites at most you should find when using either a sticky board or ether roll. When the daily mite count reaches these levels, you need to take action.

SEASON	MAX NO. MITES		
Fall	<10	<20	<20
Summer	10	20	30
Spring	5	<10	<10
	Small	Medium	Large

SIZE OF HIVE

Small: 4 frames of bees and brood in a 5-frame nuc

Medium: 10-frame deep with 8 frames of bees and brood, fully drawn comb, and a honey super partially full

Large: 2 deeps full of bees and brood, and 2 honey supers partially to completely full

Avoiding *Varroa* Mites

First, purchase strains of honey bees resistant to *Varroa* mites. Russians are very tolerant of *Varroa* populations. Not perfect, but not bad (see page 54 for more information on these bees). Several suppliers are selling strains of bees called Survivor Bees, local/regional bees, or resistant bees. The names vary but the result is the same. These strains have been selected for and have been living without chemical treatments for any malady, including *Varroa*, for some period. The longer the better, of course, but ask questions, and if the time without treatments is four or five years or more, consider them. They, too, are not perfect, but their star attribute is that they all have a genetic trait called *Varroa sensitive hygiene* (VSH). This enables workers in the colony to seek out and find capped cells of workers that have *Varroa* in with them, open the cell, and remove the pupa that is there along with the mites inside. The adult female mite may escape and find another cell, but by removing the pupa, the bees have stopped her from reproducing and removed her offspring from the equation. Sometimes the bees are too aggressive and remove the cap of an uninfested cell. If that happens they will reseal the cell. You may see some of these cells.

If you purchase Russians, Survivors, VSH, or any line that claims to have some *Varroa* resistance, you have to decide: Do you do anything to reduce mite populations in colonies headed by these bees? If you do, you reduce mite pressure, you enable these tough bees to continue as before, even stronger. If you don't, however, you are not helping find the perfect mite resistant bee.

But is that necessary? If you are not producing queens for future use in your apiary, your queen should do what she's supposed to do, and you will act accordingly to keep your bees alive. The easier it is to keep bees (because they are resistant to pests and diseases) the more likely you are to continue to keep bees. And that's where this discussion began, remember?

If you do nothing, some colonies may die from *Varroa* infestations. Nature will prune out the weakest and the worst, but not as many will die. And they won't die as fast or as often as those colonies that do not have bees resistant or tolerant to *Varroa*. It can be a tough call, but then any decision worth making is not easy.

If you choose to let nature take her course and let your bees and *Varroa* shoot it out, my work here is done. You may, in one or two years, need to replace your bees. Maybe it will be longer, and maybe not at all. If, however, you choose to lend a hand, there are things you can do that will help these tough bees.

There are a variety of screened bottom boards on the market. This plastic model needs no paint and is robust in its holding capacity. Remember, there may be several hundred pounds (or kilograms) sitting on this bottom board when there's a full honey flow. Choose one that has a sticky board option that is accessible from the rear, that is solidly built, and that has screen covering as much of the area as possible. More than an inch (2.5 cm) of support on the sides will begin to defeat the purpose of having mites drop through the screen.

Some common sense management tricks will help your colony deal with mites all the time. Using screened bottom boards is one. Some claim they actually help reduce *Varroa* populations while others say not. Either way, a screened bottom board will improve ventilation and air movement within the colony, reduce the need for fanning, allow more bees to forage, and the honey will ripen faster. (Screened bottom boards were used by beekeepers more than 150 years ago for improved ventilation. Beekeepers replaced them with solid bottom boards in the winter, and over time the expense of having two pieces of equipment fell into disfavor, and the screens went away.)

Drone Brood Trapping

Female *Varroa* mites prefer to infest drone brood just before the bees cap it over because drones take a bit longer to mature than worker larvae (fourteen and a half days versus twelve days). That tiny bit of time gives the female *Varroa* mite longer to raise one or two or as many as five more young.

One of the fundamental principles of integrated pest management: Provide an attractive bait in a good trap, let the pest wander in, and catch them in the trap. Drone brood makes an excellent trap.

A healthy honey bee colony during the active season will allow 10 to 15 percent of their brood comb space to raise drones. If you are using three medium eight-frame boxes for brood space, the queen will produce about three of those frames full of drone comb —full on both sides.

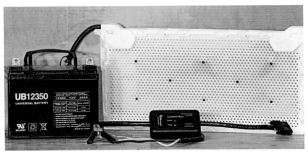

A mite zapper. Current from the battery is run through wires in the bottom of the cells. When the drone cells are capped, the battery is engaged and heats the drone larvae and Varroa, killing them. House bees remove dead larvae and mites. A safe way to control mites with very little interference to the hive.

This is what you will have if you put a frame with no foundation in your colony. The bees will fill this handy empty space with the drone comb the colony thinks it needs. You need up to three of these in your brood boxes to accommodate the 15 percent drone comb the bees want in the colony.

Keep the frames in rotation so they aren't all full at the same time and the bees always have something to work with.

If you don't provide a place for the colony to put their drone comb, they'll put it wherever they want, such as this damaged comb. It is a perfect place for Varroa to reproduce with no controls in place.

So make it easy for the queen to find drone comb. Here's how I do it.

I find the one or two frames in the broodnest that have worker foundation or comb and no, or very little, worker brood and replace them with frames with no foundation. (The worker brood I will share with another colony.) Move frames so that an empty space is created in space two or seven (one in from either side) in the middle box of your three-box broodnest. The bees will begin to fill this convenient space in that empty frame with drone comb, the queen will find and lay only drone eggs in those frames, and female *Varroa* mites will find and infest this convenient drone brood; thank you very much. Then, one week later, add a second frame in the other slot in the middle box. If your bees are building well a week later add still another frame in the third box directly over the first frame you put in below. After these three weeks, drone brood in the first frame will be mostly sealed, and the others will be in some degree of development. It'll take a season to get this going with a strong colony, but it is definitely worth the effort, even with several colonies.

Your job is to wait until much of that drone brood has been capped—meaning that most of that drone brood has female *Varroa* mites hiding inside—and destroy it, along with the mites hiding inside.

If you are beginning from a package, place one frame only for starters to see how the colony does. Because of the stress in getting started there may not be many drones raised and using three frames may be overkill. The trick is to NOT let any of the frames remain in the colony so long that adult drones emerge, releasing the mites. It is perfectly okay to put the date the frame is to be removed directly on the top bar using a permanent marker. Don't forget to remove that frame. When you have a strong colony you will be removing a frame a week.

There is a great deal of satisfaction in freezing that comb for a week then using it to replace the next one you remove. The bees will open the cell and remove the dead drone larvae and any mites. It is a lot of work, and the carnage on the front step of the colony is amazing, but knowing you have removed a large percentage of the mites in your colony without chemicals is very satisfying. You can feed the comb to chickens or birds, or melt it down for candles, but it is destructive, and using the comb over and over is much more efficient. However, the satisfaction of watching your chickens destroy all those mites is huge.

By the time you have removed the third comb a second time—that would be nine weeks after inserting the very first comb—your colony will have significantly fewer mites and maybe none at all, providing it had any to begin with.

There's a catch, of course. Drones are a sign of a healthy, well fed colony, so that should tell you something right off; if there are few or no drones during the most active part of the season, and little or no drone brood in a large colony, there's trouble afoot and you better find out what it is. A healthy, strong colony will have lots of drone brood by comparison.

Keep making mite counts about once a month all during the season to make sure they are not getting ahead of you. Keep other stresses in the colony at a minimum, and make sure plenty of food is available and the queen remains happy and productive.

With resistant bees, screened bottom boards, drone trapping, reduced stress from other pests and diseases, ample nutrition, and a good queen, *Varroa* mites should only be a minor worry in your colony management scheme. There are a significant number of beekeepers now who are doing all these things, and others who are letting go for a bit because of the difficulty of making them happen, and not treating for mites, and not losing colonies to mites. It can, and is being done.

One unknown in all of this is the level of viruses in your colony. Whether mites carry and transmit these viruses (they do for some) or simply enable them when they violate a bee's system (they do that a lot) is less important than the fact that when you have lots of *Varroa* mites in a colony, the incidence of virus symptoms in your bees is much greater. Moreover, when you see *Varroa* populations build and you enact some sort of control—whether a solid IPM program, soft, or hard chemicals—the viruses don't instantly go away because it can take from one to three months to purge the system of this scourge.

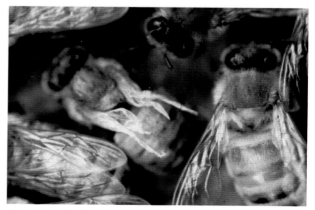

A virus that is commonly associated with Varroa *infestations is deformed wing virus. Larvae are infected when they contact the virus from house bees feeding them. The virus manifests itself by causing the wings of the emerging adult bee to be deformed. These bees either die young, or are ejected from the colony. Watch for this symptom of a heavy* Varroa *infestation.*

🐝 *Here's a trick you can use to give you an idea if there are* Varroa *present in your colony. Hold the frame so that you are looking at an angle from the top of the frame toward the bottom of the frame, so you see the bottom side of the cells. The white spots you see here are the fecal deposits* Varroa *leave when they are feeding on larvae in the cells.*

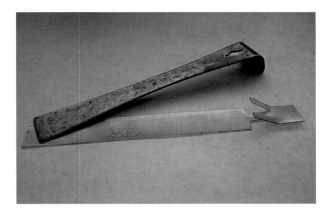

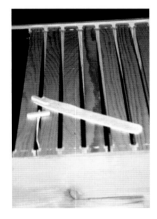

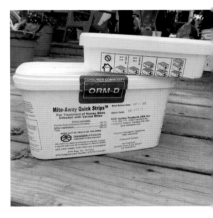

Viruses are transferred within colonies from bee to bee, from queen to egg to bee, from bee to larvae—and they will continue to plague a colony long after you have successfully reduced or eliminated the mite population. In fact, a common complaint when examining a colony that has died is that it had been treated some time ago, no mites were present in the colony when it died, yet . . . it died. Why? The most common reason is that even after the *Varroa* have been removed, the viruses remain in the worker population, taking their toll all by themselves plus adding stress to diseases such as European foulbrood, nosema, and Chalk. The colony may have died of one of those, but *Varroa* and virus killed it first.

I trust you can see now why it is far more beneficial to keep mites out of your colonies in the first place rather than to have to clean up the mess later.

🐝 *Two of the hard chemicals used to treat* Varroa *mites. Chemicals are impregnated in plastic strips and then hung between frames. Bees walk on the strips and pick up a bit of poison that kills the mites. The rust-colored strip (above, top) is an Apistan strip, the white strip (above, middle left) is CheckMite, a cumophos product. We no longer recommend using this product in a beehive although it is still legal to do so. It is extremely toxic to both bees and beekeepers, and the mite population has built up resistance to it, so it is only marginally effective. There are better products on on the market. There are two organic acids you can use for Varroa (above, middle, right). Formic acid is by far the most effective. Mite-Away Quick Strips are easy to use, but are temperature dependent. A relatively new hard chemical strip is called Apivar, (above, bottom) with Amitraz as the active ingredient. It is less of a problem for the bees than the other two, but not much. But it is effective and produces less wax residue than the other plastic strip treatments.*

Finding Safe Treatments

Let's start this off with the statement that putting poison in a beehive is just about the worst thing you can do. Everything you put into a beehive, no matter the toxicity, the form, or the length of time left in leaves a trail. All of the treatment chemicals discussed here are absorbed into the beeswax to some degree—some far more than others, but all of them are—and all of them negatively affect the bees, again, some far more than others. So do absolutely everything possible to avoid using these, and then only as a last resort to save the bees. And that is your goal: to save the bees. Do not treat any colony just because you think you should. You don't take medicine unless you are sick nor do you take it to keep from getting sick. Do not treat honey bee colonies that do not need treatment! And even then use the least toxic, least harmful, shortest-lasting treatment possible. Recall, once the treatment is over, residue remains, and your bees live in that residue as long as the contaminated wax remains in the colony. And finally, read the label and follow the label directions. Don't leave a treatment in longer than necessary (you can remove a treatment sooner than the label says, however, if conditions require removal), and do not think, "if one is good, two must be better," ever. Hold this thought: "Do no harm."

Treating *Varroa* Mites

Even the best colonies are sometimes overwhelmed by mite infestations. Either your IPM techniques haven't been successful, or a *Varroa* bomb has come home.

Let's talk about *Varroa* bombs. These occur when a nearby colony becomes overwhelmed by a mite population and the conditions are so bad the entire colony absconds, leaving their former home, looking for a better home. It is most common in late summer, after an unchecked mite population reaches the point where nearly every larva in the colony, workers, and whatever drones remain are infected with two or up to five or sometimes more mites in a cell. If you are monitoring your mite populations, your sticky board counts will reflect a huge sudden influx of mites and warn you of the problem. When mite counts skyrocket above those safe numbers on the chart you need to act fast to knock down that mite population before their viruses become endemic; before they infest any remaining drone brood (there usually isn't much this late in the season), but much of the worker brood; and before they damage a significant proportion of the adults in the colony. The adults in the colony at this time of year are the bees that take care of the bees that go into winter, and they need to be in perfect health or they will be unable to care for the winter bees who in turn will live shorter lives and perish long before spring.

This is why testing for mites is an ongoing management tool, not just a spring and fall activity. You also have to weigh the consequences of treating a colony of bees that so far seems to have held its own against mites. Can your colony cope with an event like this? Can they overcome a large influx of mites this late in the season? If they can't, then the only choice is a chemical treatment—and the softest, easiest on the bees, and least toxic choice is what a safe and sane IPM program calls for. There are only a couple of choices if this is the case.

First, let's review the hard chemicals that are available now. They are miticides embedded in plastic strips that you place in the colony. Bees walk on them and pick up a tiny bit of the chemical, which moves all over their bodies and eventually contacts a ride-along mite. That mite dies. The amount of chemical used is enough to kill a bug on a bug. But both bugs are exposed to the chemical. The poison is in the dose. Too long exposure, too much chemical, and both are injured. It's a finely tuned balance. But recall that most—as much as 80 percent, depending on the time of year—of the mites in a colony are not outside a cell riding on a bee or walking on a strip, but reproducing inside a cell. So to be effective, these chemicals need to be left in the hive until all mites have emerged and had a chance to be exposed. Constant exposure to these chemicals is very detrimental to queens in that they produce fewer eggs and die younger, and drones have their reproductive organs damaged when exposed to this stuff. In short, an inefficient, expensive quick fix ends up being a long term disaster. Hard chemicals are not a good answer.

The chemical used the longest is Apistan, using one strip for every five deep or seven medium frames of bees and brood. Honey supers must be off, and the strips remain in the hive from forty-two to fifty-six days to expose all mites to the poison.

Apivar is similar, using one strip for every five deep frames of bees and brood. It has lower residue issues than Apistan and CheckMite (this most toxic treatment available is not discussed here) and is easier on queens, brood, and drones than the other two. It stays on the same amount of time.

There is another strip, but the chemical is not nearly as toxic as the other two. It is a potassium salt of hop beta acids derived from hops, called HopGuard. The material is an oily residue smeared on cardboard strips. The bees tear the strips apart and in the process contact the chemical. When the strips are gone they need to be replaced so that the exposure time totals four weeks, and treatments need to be made three times a year to be effective.

There are two essential oil products available. One is Apilife Var, which is thymol, menthol, and eucalyptus oils soaked into a foam block. You break the block into 4 pieces and place one on each corner of the cluster in the top box. You need a shim to allow space for the foam block. These products are sensitive to ambient temperature, and need to be applied every seven to ten days three times for a treatment. Honey supers must be off, and if you apply them when the temperature is warmer, you can kill brood. Another essential oil product is ApiGuard, which is primarily thymol oil and inert ingredients. It is a gel-like compound contained in a very thin aluminum foil container. Remove the cover, place on the top bars of the top super and let it evaporate while the bees remove the gel. Don't use below 60°F (16°C) or it won't evaporate, or above 105°F (41°C) or it will evaporate too fast, driving bees out of the hive and harm brood.

The last product is called Mite Away Quick Strip. It is a gel-based formic acid product that slowly evaporates after it is applied. It is effective, leaves NO residue, but can harm queens and brood if applied when too warm, above 85°F (29°C), and is mostly ineffective if applied below 50°F (10°C). You can have honey supers on when this is working, and it's probably the best product around to deal with a *Varroa* bomb. It attacks free roaming mites and also those in cells. It is very effective. It is also very finicky relative to temperature and timing. Plus, it is downright dangerous to those who apply it. Gloves and a mask are necessary. But it is by far the most effective product there is relative to residues—there are none—and efficacy. It is very good.

As we said earlier, it's always a trade-off: efficacy versus beekeeper safety versus ease of use versus honey bee safety. If you choose to use a *Varroa* mite chemical treatment, be certain to read labels, follow instructions, and be safe—for you and your bees.

Wax Moth

Wax moths, *Galleria mellonella* (also known as greater wax moth), can be a real nuisance, but they can be taken care of pretty easily.

Sometime during your bees' first summer, a mated female wax moth will find your colony. She'll get inside by sneaking past the guard bees, usually at night. Once inside, she will lay eggs somewhere in one of the boxes with brood. The eggs hatch, and the moth caterpillars begin to feed on beeswax, pollen, honey, and even larvae and pupae, that is, unless house bees catch and remove them. If the colony is strong and healthy, the internal police force is very protective against these invaders. But small colonies, or those being stressed by other troubles, aren't as diligent and wax moth larvae can make some inroads.

Adult wax moths are about the size of an adult worker honey bee.

Wax moth larvae, like the one shown here, do great damage in a beehive. The larva is an off-white, soft-bodied caterpillar. This one is about half grown.

 As the larvae eat their way through the comb, they leave behind webbing, which hinders the bees from catching and removing them and from being able to use the comb, or to even clean it.

Cocoons of wax moth pupae are very tough, and bees usually can't remove them.

If they get a foothold, the moth larvae tunnel through your brood comb, leaving webbing and frass (excrement) everywhere in their paths. Unimpeded, wax moth larvae can, with favorable temperatures, completely consume the comb in a super in ten days to two weeks. They eventually pupate, spinning tough cocoons fastened to the sides of supers, top bars, or the inner cover. They literally chew a groove into the wood so they fit. The cocoons are so tough that bees can't remove them. Adults emerge, leave the colony, mate outside, and then the females find more colonies to infest. They'll do the same to supers you store in the basement or garage, unless you take some precautions.

You can be pretty sure there are always some wax moth eggs in your colony. Adult female moths routinely enter and lay eggs, but the tiny larvae are just as routinely removed by the bees. Thus, when finished using a super at the end of a season, those eggs are still there, but the moth police have gone.

Wax moth adults are around pretty much all year in tropical and semitropical areas, and most of the year where there are mild winters. In cold climates, wax moths are a problem only from midsummer until hard frost—outside, that is. Inside the hive is a different story.

Avoiding Wax Moths

- Don't pile too many empty supers on top of a colony; only use enough to allow the bees to keep moth larvae in check. You may find one or two larvae in a small colony, but they should be hard to find in a large colony.

- Make sure colonies are healthy. A stressed colony will have enough going on without having to deal with moth larvae.

- Do not store empty honey or brood supers in moth-friendly environments—warm, dark areas with plenty of comb to chew on, such as your basement or garage.

- Wax moth larvae do not thrive when exposed to light and fresh air. A stack of supers with a secure top piled in your basement is the perfect place for these pests' populations to explode.

- If feasible, stack unused supers on their sides with a few inches between them, and supers on top oriented at 90 degrees. This placement allows light and fresh air into the super, greatly reducing larval activity. You can also build a rack that exposes the supers while protecting them from rain. Expose, expose, and expose your supers to keep moth populations manageable.

- You can keep a super or two on strong colonies after harvest and let the bees handle the worms, if you live in moderate to cool regions.

- For a few supers, place the whole super in a freezer, set at 0° F (-18°C) for forty-eight hours. This kills eggs, larvae, pupae, and adults. Once the outside temperature goes below 40°F (5°C), the temperature essentially halts all moth activity (but does not eliminate them), and your supers are safe for the winter, no matter where or how you store them, as long as it stays that cold.

- If outside storage or freezing isn't possible, as a last resort, you can put a moth fumigant on a stack of supers. The only chemical approved for this is a formulation of paradichlorabenzene available from bee supply companies. This is the same chemical in crystal form used to protect clothes from destructive moths, but without other fragrances and additives. Do not use any other formulation of this chemical—it is a violation of the label and may damage your combs. And then use it as a last resort and only sparingly. The wax will absorb fumes. Before using next season, set frames outside one full week to air out and let as much of the material evaporate as possible.

Here is an easy-to-build storage rack for supers when they are not on a colony. This arrangement allows light and fresh air into the supers, discouraging wax moths from infesting combs. It also protects them from the elements. It is made of boards, cinder blocks, and PVC pipe, and could be made longer, taller, wider, bigger. The supers are covered with inexpensive corrugated fiberglass panels weighted with additional pipe, blocks, or other heavy material.

Storing unused supers so light and air can penetrate between frames makes them unattractive to wax moths and doesn't require a frame or shelf.

🐝 *Here is an adult hive beetle on the edge of the inner cover and another on top of the inner cover—both scurrying for cover when the outer cover was removed. You almost never see a beetle sitting still. When you open a colony they run from the light. They are about one-third the size of an adult honey bee.*

🐝 *A disposable small hive beetle trap. Fill the reservoir half full of vegetable oil, then place between two frames near the edge of the top box. Beetles will hide in the opening, only to fall into the oil.*

🐝 *Small hive beetle larvae resemble wax moth larvae, but they tunnel without webbing, and are very destructive. Pick one up and draw it between your fingertips. The cuticle is hard and unyielding, and the legs are stiff and bristle-like. Wax moth larvae are soft and pliable and easily crushed.*

Small Hive Beetle

Small hive beetles and our honey bees have had a decade or so to become acquainted since the beetles accidentally arrived in the southern United States from South Africa. They had a few years head start before beekeepers realized what the problem was, but almost immediately they became a problem—especially in small, weak colonies, and in honey houses after harvest. Beekeepers in Australia have had similar experiences and between the two countries, beekeepers have arrived at several IPM management techniques that help deal with this pest.

Small hive beetles are not an issue in South Africa because there the aggressive African honey bees manage to keep them in check. However, the more docile and inexperienced European honey bees common in South, Central, and North America were less prepared for their onslaught. And since they arrived, it's clear the beetles are primarily a tropical insect. Hives infested with these beetles can be moved from the warmer parts of the United States to the more northern areas, but generally (and I use that term carefully) both the colder winters and the less sandy soil types in the north keep populations in check. There are occasional outbreaks when beekeepers aren't paying attention, or when infested colonies move into an area to take advantage of a honey crop or for commercial pollination. When the colonies crash due to beetles, *Varroa*, or starvation, the beetles will leave and find a new home.

Beetles can fly long distances, reportedly up to nine or ten miles (14.5 to 16 km), and are attracted to colonies that have been disturbed, and have released some alarm pheromone, scientists tell us. Queen producers (in the southern region of the United States) who use small mating nucs have problems

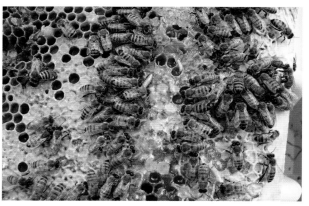

🐝 *As the larvae tunnel through the comb they defecate, which causes the honey to ferment and become runny and is fondly called slime. In extreme cases the slime will run right out the front door. The bees will abandon the hive long before it gets to that point.*

because there aren't enough bees to defend the colony, and very often colonies that go queenless decline and are destroyed by the beetles.

Even in southern United States winter takes a toll on these beetles and their populations don't build until midsummer or so. It's unlikely you will see beetles in your package colony— perhaps a very few early on that came with the package. But if you do see them early, you will likely see them in greater numbers as the summer moves along. You will notice them especially on the top of the inner cover when you remove the cover, or on solid bottom boards. Generally, they hang out in locations in the hive that are isolated or protected from the bees. Bees will harass them, sometimes to such a degree that they will imprison them within walls of propolis on top bars and other locations, like the top of the inner cover, to keep them from roaming and destroying.

Destruction can be extensive. Mated females lay clusters of tiny eggs in locations the bees don't have access to. The eggs hatch and hundreds, even thousands of aggressive larvae tunnel through brood comb, honeycomb, any comb seeking pollen and especially larvae and pupae to eat. As they tunnel they excrete a liquid mess that causes any honey it contacts to ferment and bubble and turn into what beekeepers call *beetle slime*.

When the beetles attack small, weak, or disorganized colonies the resistance put up by the bees is weak and futile. Soon the beetles command much of the unoccupied portions of the colony, then in mass invade and destroy the remainder. Colonies that are strong with larger populations can often hold their own against an invasion, and with a little help from the beekeeper the bees can win this battle without using harsh chemicals or poisons.

Avoiding Small Hive Beetles

The first line of defense against small hive beetles is to maintain large, strong colonies that are not under stress from other pests, diseases, or nutritional issues.

If you have weak or small colonies, especially those splits made later in the summer, there are several traps available that are effective in keeping beetle populations down. All offer the beetles a place to hide from the bees, and once in, offer no way out. The differences are mainly placement in the hive, size, and cost.

Some are made like trays that sit on, or replace the bottom board of the hive. There are slots in the top the beetles can enter but the bees can't, and the bottom tray holds a supply of food-grade vegetable oil the beetles drown in. Though effective be aware that they replace the screened bottom board in the colony and that the colony must be nearly level so the oil doesn't run out or collect on one side or the other.

Tray traps sit on the bottom of hives and work much like the sticky board for Varroa. An oil-filled tray awaits the beetles beneath the screen. If you have a small number of colonies you may want to consider these as they are quite effective.

You also will need yet another large piece of equipment. Other versions of this trap are made so that the tray holding the oil can be slid in from the back, fitting into a slot in a specially made bottom board that could be screened. With this version you needed fewer trays and could still have screened bottoms on the hives. It was a better move. There are several similar models available that offer mostly adequate control, but it is another piece of equipment.

There are several trap models available that fit between the top bars and sit just below the surface of the top bars so a beetle running from bees will seek out this hiding place. Of course, the holes provided only lead to a oil-filled trap below. These are effective, inexpensive, and reusable. And you can use several in a hive at the same time.

The small hive beetle life cycle requires that the larvae, when ready to pupate, leave the colony and burrow into the ground outside to complete their metamorphosis. There is a ground spray that can be applied to the area around your colonies that will kill the larvae when they cross it. It also is effective against fire ants, but it must be reapplied after it rains. If you routinely move colonies it doesn't do much good either. Trapping seems to be the best method of dealing with beetles once you have them in a colony.

Small Hive Beetles and Honey Supers

Moving small hive beetles into a honey house after harvesting honey supers can be a disaster if left to their own devices for a few days. Without interference from bees any beetle larvae in the supers can and will run amok in the supers, ruining all of the honey either by tunneling through it, or having the slime run down and over frames they didn't touch. You have a short window of time between moving bees into a honey house situation (it may be your garage or basement or kitchen) and extracting before they begin to do significant damage. If you harvest but must store for a few days and you know or suspect beetle larvae are in the supers, run a dehumidifier in the storage room as much as possible. This will help dry any uncapped honey you inadvertently harvested, and the drier air dehydrates many of the beetle eggs and tiny larvae that may be present.

This may not be feasible, so the best bet is to always extract your honey just as soon as you can after harvesting. Any adults or larvae will be removed from the finished honey in the process and if any larvae remain in the honey supers, the bees will clean them out when you put your supers back on the hive.

Small hive beetles are no longer the scourge they were initially, but you do need to pay attention to them if you live in the warmer areas of the United States and Australia, if you run small or weak colonies, and when harvesting.

Animal Pests

If your colony is raised off the ground a couple of feet, skunks and opossums are seldom a problem, but continue to check for them. Skunk visits are noted by torn-up sod or mulch directly in front of the hive, and muddy paw prints or scratches on the landing board. A skunk will, if it is reachable, scratch at the hive entrance at night. Guards who investigate are slapped, grabbed, and eaten. Other guards will fly and sting the intruder, but skunks are nearly immune to stings on their paws, face, and even inside the mouth.

This can last for some time during a night, and for many nights in a row. A mother skunk will bring her kits and show them how to harvest this sweet, high-protein snack. A colony that's attacked will become very defensive because of the constant disruption and continuous exposure to alarm pheromone, especially the day following the attack. Opossums are generally opportunists and grab what they can without the scratching.

Raccoons may also investigate your colony. They are usually attracted to a hive because of wax or propolis carelessly discarded in the vicinity of the hive. Here's a word to the wise: Raccoons don't attack the front door. They will, if determined, remove the cover and inner cover (still loose from your recent inspection) and pull out a frame. They'll drop it to the ground, and then drag it away several feet. Guard bees or any bees on the frame fly or crawl back to the hive, leaving the raccoon to feast on honey and brood in relative peace. Placing a brick on top of your hive satisfactorily prevents this.

A piece of cut comb in a plastic container ready for sale: With this you don't need uncappers, extractors, bottles, or pails, and you get to eat honey at its very best.

Comb Honey and Cut-Comb Honey

There are some options when deciding what kind of crop you would like your bees to produce. It may not be just honey in a jar.

Cut-Comb Honey

Many beekeepers get into beekeeping because of the bees, and get out of it because of the honey (and now, *Varroa*). If honey is not the goal for your beekeeping, you can always let another beekeeper harvest the surplus honey your colony produces. This is easy, inexpensive, and satisfying: The beekeeper gets paid in honey, or money, and you can get enough to share with friends and family. If you are interested in producing some honey and there's no one to help with extracting and purchasing uncapping equipment (no matter how basic), there are a couple of other choices.

The easiest is to make cut-comb honey. You need to make only minor management and equipment changes to produce it. Special beeswax foundation, nearly paper thin and easy to cut, is used in your frames instead of plastic foundation or regular beeswax foundation with support wires in the wax.

There should be a good, ongoing honey flow when these supers are added so the bees draw out the foundation and fill the cells with honey rapidly. Once filled and capped the crop is ready to harvest. Remove these boxes using the same techniques as with any other honey super.

This honey is not extracted, however. No uncapping, no extracting, and no jars or pails. Rather, remove the frames from the super, place them on a draining tray, and cut the honey comb into several pieces. A draining tray is simply any tray that can be washed, with a piece of wire mesh on top of it. Before cutting, obtain some of the clear plastic boxes that bee supply companies sell for this purpose so you'll know what size to make the cuts.

When the bees have filled and capped the frame, remove it from the colony. Once harvested, place it on a drainage tray (a tray overlaid with a grid, such as mesh screen). Cut the comb into pieces that will fit into your containers. Let drain overnight, then place in your containers (available from bee supply companies) for this purpose.

The parts and pieces of a Ross Round Comb Honey Frame: To use, insert the sheet of thin foundation, put the rings in the holes and snap the two sides of the frame together. Eight frames fit in a ten-frame super (six in an eight frame), and there are four round sections in each frame. To harvest, open the frame, lift out the rounds, trim the excess foundation, cover, and you have your finished product.

Cut the comb into correct-size squares and let the honey from the cut edges drain overnight so they are dry the next day. Use a plastic spatula to pick up and place the squares into the plastic boxes. Package and distribute to friends and family. Use cut-comb honey as is by the spoonful; put a spoonful on hot toast or biscuits, or anywhere you want a wonderful honey taste, exactly as the bees made it, untouched by human hands. And yes, you can swallow the wax.

Comb honey equipment leads the pack in innovation of design and ease of use. Years ahead of the traditional beekeeping equipment producers, comb honey producers developed two-piece frames with only a single sheet of foundation added (no wires, wedges, slots, or pins required). With these, when the bees finished filling the spaces, the supers containing the plastic frames are removed, and the frames are divided into two. What remains is a round section of comb honey, capped by the bees, already in its container, and ready to eat. All you have to do is cover each side.

Ross Round sections: showing wet (left) and dry (right) cappings, ready to sell. Both are excellent, but dry cappings always seem to sell best.

Using Cut-Comb Frames and Foundation

Any frame that fits your supers can be used to make cut-comb honey, but try to purchase frames from the supplier that are made for your supers. These will usually have the best fit. However, cut-comb frames are in the hive for a short time so spacing issues will be minor.

For cut-comb production you don't need foundation embedded with wire, nor do you need to add wire to the frame for additional support. You will need to use the wedge on the top bar though, which can be a challenge. You do want to use cut-comb foundation, however; it is thin and easy to cut, and it isn't tough when you eat it along with the honey.

Once you have frames and foundation ready, put three or four in the center of your honey super, but for the remaining frames use regular, wired foundation. I suggest using a queen excluder to make certain the queen doesn't come up and put brood in these frames, which makes the frame unusable for cut-comb honey.

Comb Honey History

There's a wonderful history associated with comb honey equipment and the legacy of producing this perfect crop. When the movable frame hive came along in the 1850s producing extracted, liquid honey blossomed. Sadly, taking advantage of this product blossomed too—unscrupulous honey packers could add inexpensive sugar syrup to the honey, diluting its cost and quality, but not its texture or thickness. As a result consumers began to doubt the honesty of liquid honey producers, favoring comb honey even more since it seemed impossible to tamper with its quality.

Comb honey's popularity exploded and manufacturing the basswood (*Tilia* sp.) equipment exploded also. Basswood's long grain enabled it to be easily bent to form the wooden sections that held the foundation the bees would fill with honey comb, then honey. These were harvested as is, complete with cappings intact—the most perfect form of honey imaginable. These wooden boxes were mostly produced in the northern parts of the United States because the natural honey flow was fast and furious, which was necessary for filling those basswood boxes.

In Medina, Ohio, the A. I. Root Company made millions and millions of these basswood sections. And after a very few years their sources of basswood began to grow thin. To maintain a steady source of the wood, they planted plantations of these trees.

After only a very few years the basswood trees were large enough to provide shade for the Root Company's queen-producing beehives. Soon, the trees were large enough to bloom, and during the warmest part of the summer, these miracle basswoods not only provided shade for man and bees alike, they provided a most wonderful nectar—a great and glorious crop of basswood honey in only a very few weeks.

As the trees were thinned they provided even more wood for the basswood sections. Basswood honey, made from the flowers of basswood trees and put into the crisp, clean basswood sections was a rare treat.

The A. I. Root Company's basswood trees made shade, made honey, and eventually made basswood section boxes, the precursor to comb honey production.

Summertime Chores

By early summer, your brand new package colony has most likely grown past the rigors of becoming established. The stresses of temperature fluctuations are reduced, the population is building, the queen is producing, nectar and pollen are being collected, frames are mostly built up with beeswax cells, and honey is being stored. This is the norm.

But if you have a dozen colonies, one or two may not be thriving and need some additional attention. Having at least two colonies is recommended, so you'll have a basis for comparison. How can you tell what's happening, no matter where you are? The bees don't send out press releases, but their activities are indicative of their situation. Mostly bees react to their environment rather than plan ahead. You should, however, be a step ahead of them so you can anticipate their needs. Temperature extremes, rainfall, and even weed growth are limiting factors for your garden plants, and for the plants your bees visit. But even with that experience, an occasional chat with nearby beekeepers can be enlightening, and belonging to a local club becomes even more valuable. An experienced local beekeeper can, in a few moments, share the typical season's progression—that bit of wisdom is worth its weight in gold.

Surplus

Surplus is the term beekeepers use for the honey they harvest. It is anything more than the 40 to 60 pounds (18.1 to 27.2. kg) the bees need to overwinter. Your package colony—playing catch-up because of having no stored honey, having to build the entire wax comb, and raising young—starts the season essentially at a deficit.

To store surplus honey, your colony first needs to produce the equivalent of about 60 pounds (27.2 kg) of honey plus continue to feed a growing population of young bees. Your package bees have to really hustle just to stay even. And, if the season's off they may never get even, and you'll need to feed them most of the summer. That has become increasingly common because of iffy queens, changing weather patterns, and *Varroa* pressure. Don't let a colony starve in the midst of plenty because you weren't paying attention. Next season, they'll have a reserve of food to feed a growing population and won't have to produce nearly as much wax.

At the same time, keep your record book up to date. Your bees are exploring their environment and finding, or not finding, sources of food. Until you know as much as your bees do, you're still learning.

Your advantage, of course, is that you can keep records. A colony of bees has to reinvent the wheel every season. The few bees that overwinter were born in the fall and have no institutional memory, as it were, so the colony as a unit must learn it all again every year. You, on the other hand, can quickly review last year's records and recall flowering times for fruiting plants. Being aware of how the season progresses, whether short and ending early, or the longer season of moderate climates allows you to anticipate swarming, honey flows, and preparation for the slow or winter season. Your package colony will be playing catch-up most of its first season, but it will be reacting to the local environment in much the same way as more established colonies.

Your package colony, in eight to ten weeks after installation, will probably have brood in all three of their brood chambers, but that will range from a box and a half to three boxes full. The difference will be dependent on how favorable the weather has been, the type of bees you have, and any obstacles that have arisen during that time.

By midsummer, honey will probably be in at least one and perhaps part of a second honey super. Looking ahead, you must prepare for the slow season—cold weather in the cold regions, and the rainy or cool season in the warmer areas—and make sure your colony has enough honey stored to last until spring returns.

In the coldest regions, where snow is common and winter lasts six months or more, a typical colony will need about 60 pounds (27.2 kg) of honey to survive. An eight-frame medium super will hold, if completely full, roughly 30 to 40 pounds (13.6 to 18.1 kg) of honey. In the warmer areas, 40 pounds (18.1 kg) is about all you'll need, and in the warmest areas, where forage is available year round—southern Florida, for instance—additional food is not required.

By midsummer, honey becomes an issue for you and your colony. How much do you want? And how do you manage that? Certainly nature may make that decision for you, but during an average season, your package colony may make 40 pounds (18.1 kg) of surplus honey. It can make more in a good year, but maybe none during a poor year.

You must provide room for honey production—room to dehydrate nectar—using an additional super or two. Deciding how many supers to provide is definitely an art. If you provide too much room, the super will remain unused, which isn't too great a problem the first year.

(Continued, page 138)

Finding Queens, and Requeening

During the summer you may find your queen missing. This may be from being poorly mated, nosema, being exposed to harsh chemicals, or other problems. Alternatively, the queen may be injured while you are working a colony, and a damaged queen is damaged goods. Or, perhaps her performance isn't up to par—spotty brood pattern, not much brood, too many drones—or she just quit laying. After checking with a beekeeper with more experience than you for advice you may decide, for whatever reason, to requeen.

Order a marked queen from a local supplier, or mail order. Prepare the colony for when she arrives. If the colony has a queen, but a bad one, you have to get rid of her first, and now comes the most dreaded phrase in all of beekeeping: First, find the queen.

If the colony's population is small, finding the queen may be easy. Here are a couple of tricks that may help. The first one is labor intensive and time consuming, but it generally works.

Have an extra box, and three or four large boards or extra covers with you to start. First, put the extra super on one board. If you have two or three brood boxes, lay down anther flat board (the cover if it is flat) and put the top box on that. You don't want bees falling out the bottom. Lay another flat board on the top box and put the second box on that, if there

is one. You will start with the bottom box on the bottom board. Have the empty super close by with a bottom board or regular bottom right on the hive stand, if you can. Remove the nearest frame. Look at the next frame as you lift to see if she is running from the light. Examine that frame carefully and if not there, place it in the empty super. Use smoke sparingly.

Continue looking for the queen. When the first box is done, move the next one, on its board, on top of the now-full super and examine it frame by frame, putting examined frames in the box you just emptied. Eventually, you should find the queen.

If you don't find her the first day, you can try the second tip, and that is to put a queen excluder between the bottom and second box. When you return in three days, where the eggs are is where the queen is.

Once you find her remove her from the colony. Before she is dispatched to honey bee heaven, ask if someone wants her as an observation hive queen, or to hold a nuc until a mail order queen is ready, or for various reasons.

When the queen is gone, a new queen should be introduced within twenty-four hours. Put the new queen cage, with a queen and the attendants in the colony the same way you did at first, but now wax comb will help hold it in place. (See pages 94 to 96 and 99 for complete instructions for introducing a new queen.)

With drawn comb in the colony you can simply push the new queen cage into the comb and have the wax hold it in place, candy side up.

Laying Worker Colony

Though uncommon, bees in a colony that have gone without a queen for a while may begin to lay eggs themselves. When examining the colony you may see some brood present where there wasn't any a week ago. Usually the brood will be scattered over the surface of several frames, with no order or plan to where it is put. If it is there long enough—six days—the bees will cap it with large, bulging wax coverings. This is a sure sign of a laying worker colony because those are drone cell cappings and workers can only produce drones. Now you have a different problem.

You must replace the lost queen as soon as possible. And you should know that once a colony has reached this point, recovery isn't guaranteed.

If you started another package that is progressing normally, take the fullest frame of brood (complete with eggs and open and sealed brood), brush off the bees, and put the frame into the colony with the new queen cage. If available, take two frames from this other colony and add them to the distressed colony. Put the new queen's cage adjacent to the brood frame or between brood frames so the bees in the colony get the essence of the eggs, open and sealed brood, and a queen.

This may halt the development of additional laying workers, but it probably won't stop those that are laying. Your only hope is that they will soon expire leaving the colony once more with only one laying female—your new queen.

If the queen is accepted, she will slowly bring order to the colony, begin to lay eggs, produce colony-cohesive pheromones, and life will return to normal. This is the most likely scenario.

If this queen is not accepted and you have no other to draw on, this colony is most likely doomed. You can try to purchase a summer split from a local supplier. Your last choice is to get rid of the doomed bees by shaking them in front of your other colony and put away your equipment for this season, ready to try again next year.

This is an example of a good brood pattern and a productive queen. This is what you are looking for by mid- to late summer in most places: lots of bees, and a fair amount of open and sealed brood in the colony.

If you discover a pattern that is as disorganized as this is, with drone brood in worker cells, the colony is in trouble. It may be a laying worker colony, or a drone-laying queen. In either case, if there's a queen she must be found and removed and a new one added, and if there isn't a queen a new one must be added.

A frame loaded with queen cells (on the face of the frame rather than hanging from the bottom) usually means the colony has lost a queen and has constructed emergency queen cells to replace her.

Your bees will store some honey in the broodnest boxes. You want them to store enough but not so much they fill space the queen needs to continue producing the next generation of workers. In fact you must prevent this by moving frames full of capped honey out of the broodnest and up into the honey storage supers.

To get the necessary amount stored, you want that 40 pounds (18.1 kg) in one super and an additional 20 pounds (9.1 kg) stored around the edges of the broodnest. That's about eight full frames of honey out of the twenty-four you have in the three brood boxes. If there is more, you should move some out. If there is less, you either move some in or reduce available storage above the queen excluder.

If you are having a bumper crop, you need to continue to add boxes, two or three, as the first boxes added are filled. Here's a trick if your colony fills those bottom three boxes with almost all brood and stored pollen and keeps putting honey above their excluder. Take a nearly full, partially capped honey super and place it on the very bottom of the colony. This is an abnormal situation for the bees, and they will actually (almost always) move it up into the broodnest. This solves two problems: not enough honey in the broodnest for overwintering and not too much above. During the summer, don't get too carried away adjusting honey storage. Usually the bees figure it all out without too much interference, but keep an eye on it.

Late Summer Harvest

Your first-year colony probably won't have a harvestable amount of honey before late summer, if then. The conventional procedures for harvesting are outlined later and are simple enough to carry out alone, and even easier with help.

But now, preparations for winter begin in earnest, and they need attention. As day length shortens and weather cools, the queen's productivity slows and there will be less brood. Your drone frame may be mostly empty, but keep it in the hive until you harvest, replacing it at season's end with a full frame of honey for overwinter stores. Check the broodnest very carefully for signs of disease, because this is when problems you may have missed earlier begin to emerge in a first-year colony.

There will probably be incoming nectar for a bit, but examine closely the amount of stored honey, making sure there's at least the 60 pounds (27.2 kg) needed for the slow season. Be mindful that a first-year colony may need additional feeding to supplement that ration and be prepared with feeders and sugar.

Fume boards are easy to make. This one is made from scrap pieces of 1" x 2" (2.5 x 5.1 cm) lumber and plywood. The inside is lined with something as simple as thick, absorbent cardboard. You can purchase preassembled models that come with cloth liners.

Removing Honey

If at the end of the season, before fall sets in and before you apply any necessary medications, your colony has surplus honey, it's time to harvest. Countless beekeeping books explain nearly countless ways to do this, but for a hobby beekeeper with only a very few hives, neighbors, and not much time, there's only one good way to do this.

Using a Fume Board

To remove a box with frames full of capped honey and leave the bees unagitated in the colony below, you must use a fume board. This simple device looks much like a telescoping outer cover, except it is the exact outside dimensions of a super, not larger, so it sits on top of the super below it. You can make or buy one. The principle behind this technique is fundamental. On the inside of the fume board is an absorbent pad. It can be flannel, cardboard, or wood. You apply the correct amount of a chemical repellent to the pad, put the fume board on top of the super full of honey (cover and inner cover removed) you

🐝 *This frame has brood in the center and honey on the edges and sides and top. The honey stored here will feed brood during the winter.*

want to remove, and wait ten to fifteen minutes for the bees to move down into the colony away from the fumes and out of the honey super you want to remove.

There are several products available to use with a fume board. They are all essential oil based, and although you and I would find them almost appealing, the bees don't like them at all and will move away from them in the hive—downward. These chemicals aren't absorbed by the wax or the honey and are nontoxic to the bees. Fume boards are usually dark colored or black (furnished that way or painted by you), and on a sunny day they will warm the repellent, vaporizing it, and flushing the bees out of the super. In a few moments the bees will be gone, and you can remove the bee-free super and place it in a bee-safe location.

The reason this fume board works so well is that your neighbors will not even suspect you are harvesting. If correctly managed, no bees fly, no bees become defensive, and no bees even seem concerned. However, too much of a good thing can and will cause a problem. Apply just the small amount of repellent recommended by the label. If you overapply, you will chase bees out of one, two, maybe three supers. And they'll run right outside. Suddenly you will have a flood of bees pouring out the door, stumbling over each other to escape. This is not good. Err on the side of not quite enough at first. You can add more if needed, but you can't push bees back inside the hive.

You can't just leave surplus honey in the colony over winter. It will most likely crystallize in the cells. Cool temperatures hasten crystallization, and those frames will sit on a colony, uncovered by the cluster, unprotected from wax moth, and turn hard as a rock.

When this happens, you can't get the honey out, no matter what you do, and neither can the bees. Come spring it's still there, and neither you nor the bees can use the frame.

Your bees need some honey, certainly. Bees will consume 40 pounds (18.1 kg) if it's warm all winter, maybe as much as 60 pounds (27.2 kg). A medium frame full of honey has about 5 pounds (2.3 kg) of available food. To supply 60 pounds (27.2 kg), use 12 full frames of honey. Often, but not often enough to gamble on, a first-year colony will have six frames of honey somewhere in the three brood boxes. It may be 12 half-frames or some other combination. Examine them so you know. That's 30 pounds (13.6 kg) or so. There may be more, so look carefully.

A full super of honey frames added to that will give your bees 70 pounds (31.8 kg). Or, they may have more below and need less above. This arithmetic is necessary.

Don't guess about the amount of honey to leave the bees. Too little and they will starve in late winter, when the need for lots of food to feed lots of new brood escalates. In fact, late winter is the time of year that most colonies perish—either because they simply run out of food, or workers raised in the late summer were damaged by *Varroa* mites earlier in the season and died young, or, more likely, they simply fell victim to one of the many viruses passed along by other bees in the hive. This early demise is one of the contributors to the malady labeled Colony Collapse Disorder.

Several things influence the actual date you will remove honey from your colony in preparation for the coming fall and winter seasons. Foremost of these is the need to treat for *Varroa* mites. If you are using any of the hard chemicals or the essential-oil potions, you must not expose honey that's intended for human consumption to these chemicals.

Beware of Foul Fumigants

There are other chemical repellents that work— Bee-Go and Honey Robber are their names—but think twice before using them. They are toxic, flammable, and the foulest smelling concoctions ever created. They are effective, efficient, and will not taint honey or equipment. Spill them on your bee suit, in your car, or in the house, however, and you will be sorry forever or until you move or destroy the fouled items. These are commonly used by commercial beekeepers who are skilled in handling them and have the necessary equipment to store them.

So, treating those bees raised in the fall for *Varroa* mite infestations influences when to harvest honey. In order to ensure you have healthy bees going into winter, remember that you need to take care of the bees that take care of the bees that go into winter. Thus, in northern areas, you need to consider treatment IN JULY! Not August, or later. JULY.

Another consideration for harvest time is your time. If you are removing honey and giving it away, your time commitment will be minimal—perhaps an hour on one weekend day. However, if you'll need to prepare a space—the garage or the basement—harvest and move those frames there, uncap, extract, remove the frames from the area, and clean up. It can take the better part of a couple days with one or two colonies. Plan accordingly, and factor in when treatments will need to begin. Your record book will help here, especially next year.

Removing Frames or a Super

The best advice for learning to harvest honey is to help someone else do it before you have to do yours. But that may not be possible, so here are some hints and tips for making this task as easy as possible.

If time permits, the day before you harvest, quickly examine just the honey supers you will be working in the next day. Smoke the bees a tiny bit so they leave the super and go below. Then quickly loosen every frame with your hive tool, breaking all bridge and brace comb and other anomalies. Loosen, too, the whole super so you can move it easily the next day. Overnight the bees will clean up any honey from the broken comb and you won't have leaky, sticky frames to handle and store.

This quick exam also shows just what you'll be harvesting, so you have the right equipment ready. If the super you have

Get all your equipment together before you begin. Use a plastic, tote-type container to transport individual frames of honey from the beehive to wherever it is that they will be processed or packaged. Remember, honey is a food and should be treated with respect. Keep everything it touches clean and tidy. Your family may be eating it tomorrow.

has surplus honey in it (additional to the amount needed by the bees), whether completely full or with only a few frames, here's how to remove them quickly, easily, and safely.

- Assemble everything you'll need, check the weather and your neighbor's activity, and if all's clear, get everything to the colony.
- If you'll be pulling only a few frames, put your container next to your colony, with the lid off and nearby. Have frames ready to replace those you remove.
- If removing a whole super, lay down the extra cover, board, or sheet of plastic and have the top ready, on the hive stand or in the cart.
- Prepare the fume board.
- Squirt or spray just barely enough of the repellent on the absorbent material inside.
- Err on the side of too little rather than too much—you can add more, if necessary.
- Don't smoke the front door.
- Do remove the cover, smoke, wait, remove the inner cover, and smoke again. You want the bees to begin moving down to the super below.
- Waft two or three puffs over the top, and place the fume board on top.
- Wait.
- Wait a little longer.
- Check beneath the fume board to see if there are bees there.
- If not, gently lift the super and look at the bottom. If you see bees, replace the fume board and wait a bit more.

Harvest Equipment You'll Need

- Fume board and repellent. Pick a warm, sunny day if at all possible. Make sure your repellent container is open and unsealed before you need it.
- Smoker, hive tool, protective gear, extra frames, and *Varroa* treatments, if using.
- Bee-tight container to put frames in—a plastic container large enough to hold as many frames as you'll need (up to eight) with a tight, sealable lid.
- If removing an entire box, a bee-tight bottom and cover—two covers work if you have them, but two pieces of plywood or heavy-duty plastic sheeting will also work.
- If the supers are too heavy or too far to carry, you'll need a cart.

- If after ten or fifteen minutes there are still bees on the bottom, add a tiny bit more repellent to the fume board.
- Wait.
- When the bees have cleared the super, remove the frames to be harvested or the whole super. Leave frames with unripe honey. (Ripe honey is covered with wax cappings.)
- Put full frames in the container and cover between adding another frame. Don't start a robbing melee!
- Replace frames removed with empty frames. Don't leave an empty spot, even for a few days. Put frames with foundation only on the edges, moving frames with comb to the middle.
- If removing a whole super, place it on the board or plastic and cover it immediately.
- Apply *Varroa* treatment, if planned.
- Close up the colony.
- Move the honey and gear inside, and pat yourself on the back.

A good place to store a few frames is in the freezer if you aren't going to deal with them immediately. Otherwise, get them to the place they will be handled as soon as possible. If you are disposing of surplus honey (uncommon, but don't count it out) double-bag it so it won't be robbed and put a lid on the trashcan. All manner of animals will be interested if you aren't. A better solution is to simply give it to another beekeeper to extract. He or she will either keep or share the honey. Either way, the task is resolved.

Harvesting Honey

Bees store honey. During a good year, they'll fill every space you provide. Honey is the primary reason many people have bees. Here's where you have to take an honest look at this part of having bees. It reflects back to the zucchini complex mentioned earlier, but it applies to any garden harvest. How much honey can your family eat or use?

Honey yields vary every year. The amount of rain, varmints ransacking the place, temperatures that are too cold or too hot, too little attention—all are factors that affect how much or how little honey you'll get. But most years, your colony will produce between 40 and 60 pounds (18.1 to 27.2 kg) of harvestable honey. Some years, with adequate attention, 100 pounds (45.4 kg) isn't out of the question.

How much honey is that? A typical 5-gallon (19 L) pail holds 60 pounds (27.2 kg) of honey. That's not a lot to distribute. If you're raising comb honey, that's 10 or 12 round comb frames, or 120 or 130 of the small rectangular containers.

If you are realistic about how much to expect before you begin, you can plan the type of honey to produce, how to best manage for production, and how to process it and handle it once harvested. This goes a long way in avoiding the zucchini complex.

You'll find frames that have what are called "wet" cappings, as shown at top, and "dry" cappings, at bottom. When bees place the wax covering over the cell filled with ripe honey, they either place the wax capping directly on the honey, giving the cap a wet appearance, or they leave a tiny airspace between the wax and the surface of the honey, giving the cap a dry appearance. Comb honey producers prefer the dry look, but neither wet nor dry caps have any effect on the quality or flavor of the honey.

Extracted Honey

This is the most common form of honey, the one you're already familiar with—the kind sold in a jar.

To produce liquid honey, you give the bees frames with plastic foundation. They build beeswax comb on the foundation, fill the cells with honey, and cover them when the honey is less than 18 percent water (called ripe honey).

Then, you remove the frames, and remove the wax cappings, leaving almost the entire honey-filled comb intact on the frame. You'll save the beeswax and honey you remove for later use.

The honey-filled frames are then put into a machine, called an extractor, which is powered by an electric motor or muscle power. The extractor spins the frames at a fairly high speed, so the liquid honey runs out of the cells and is collected in the bottom of the extractor. It works just like a salad spinner, except that you drain the honey into jars or pails. The frames are given back to the bees for a time, so that they can clean up the sticky bits left in the box.

🐝 *Uncapping knives range from heat-controlled units to heated units without controls to unheated uncapping knives (uncapping knives have offset handles so you don't bump your knuckles on the frame as you remove the wax) to simple serrated kitchen knives, which work well for processing twenty to fifty frames.*

🐝 *Cut the cappings off using a knife, and let the cappings fall into the tub.*

🐝 *A practical uncapping tub has a large container on top to catch cappings and hold uncapped frames until ready for the extractor, a metal grid to support the cappings, and a holding tank below into which honey can drain. A valve drains honey from the bottom tank. A removable mesh liner filters the honey before it drains below so you don't have to filter it again.*

🐝 *A hand-powered extractor holds four frames at a time.*

🐝 *An extractor with a small motor makes the task much easier.*

An uncapped frame, with honey ready to be removed.

Drain the honey from the extractor and strain it to remove bits of wax, propolis, the errant honey bee, and the like. Many types of strainers are available. Most fit right on 5-gallon (19 L) pails. Honey can be conveniently stored in reusable, clean 5-gallon (19 L) pails with lids. You can't have enough 5-gallon (19 L) pails if you are in the honey business.

Extraction can be done whenever there is ripe honey to be removed. Most people who extract honey tend to bunch their activity so that the setup, extraction, and cleanup time is efficient. Also, larger extractors need a minimum number of frames to be full and to run.

One way to handle this task is to take your supers to another beekeeper who extracts honey at the same time you do and combine your frames to fill the extractor. All manner of negotiations occur for this service, because that's what it is—a service. Assisting with the work is a good way to learn the task without buying the equipment first, and two always work faster than one.

Leaving some amount of honey may be part of the deal, along with the cappings wax, or there may be a simple per-pound charge for the task. Often you can purchase beeswax that has been cleaned and melted, avoiding that task also. Of course it's not your wax, but beeswax is beeswax, after all. This isn't a free service, so don't expect that. But with only two or three colonies, justifying the cost of all the equipment needed to extract and process this small amount of honey and wax can be difficult.

Even if you can't locate someone to do this, visit a beekeeper or two who already have the equipment. The whole process is complicated enough that a bit of study is warranted before you begin. Take notes and pictures.

Once you have a feel for the workflow design and the equipment you'll need, where will you set up all this at your house? A basement is often used, because it's inside, away from the elements and the bees, out of the way of others, and there's running water to clean up with. A walk-in basement can work well because you're not carrying supers down and then back up stairs.

Kitchens are, or should be, out of the question. Wax, honey, propolis, and bees in the house and on the floor seldom make good impressions.

The garage, as long as it's relatively bee-proof, is often used, and you can clean it later.

The first equipment you need to consider is a way to protect the floor, or better, a floor with a drain, which allows you to thoroughly wash the floor after extraction.

Supers with frames full of honey are brought in and set on some sort of catch tray—an overturned telescoping cover works—to keep drips from running out.

If you're doing the extraction work alone, think through the process. Cappings are removed using a variety of tools but most commonly a type of large knife. For only a few frames (two or three supers), a serrated bread knife is usually sufficient. For more frames, a specially designed knife should be used, and for numerous frames (ten or more supers), a heated knife is best.

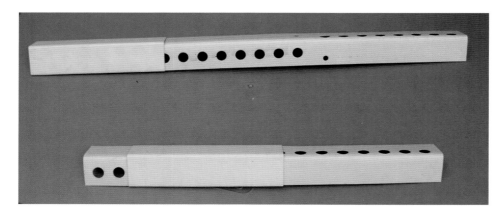

Mice can't chew through an expandable, fits-all-hives guard such as this one. It's the best there is.

The cappings are cut or scraped off into a tub or specially designed tank. Then the frames are placed in the extractor, spun, removed, and put back in the supers. Each step needs a plan. Where do the frames go after they are uncapped but before going into the extractor? They'll fit into one of those tubs, if you have one. What if two people are working together? Is there room? Can the equipment handle double speed?

When planning your workflow, remember the very true, very sad observation made years ago by Ohio's Extension Specialist for Apiculture, Dr. James E. Tew: "Most people get into beekeeping because of their curiosity about bees, but they leave beekeeping because of the nightmare of harvesting honey." *Nightmare* may be a bit harsh, but without planning, it can be a headache.

Fall and Winter Management

When you've finished extracting, put the supers or frames you extracted back on the colony and let the bees reclaim the sticky bits left behind. After a day or two, leaving the cleaned frames in the super, remove the whole super used for storing surplus honey. This leaves the three brood boxes, where your bees will spend the next few months.

Pest and Predators

Medications and treatments for *Varroa* mites have evolved since they first arrived, and since viruses became more of a problem. Since you have been monitoring mite populations all summer, fall treatments should not be necessary. However, *Varroa* bombs or other issues may lead to too many mites and a treatment may be required. Most mites now will be adults, exposed because there isn't much brood to hide in, so a quick treatment should work. If absolutely necessary, use a formic acid treatment for mites now as it will knock down adults, and if left on for only a week, or even just a few days, will damage many of the mites in cells also. This is not an ideal way to do

this, but time is not on your side now, and the goal is to have your colony survive.

Check also for diseases, the queen, brood pattern (there may be almost none this time of year, but look closely), and honey stores. Recall the 60 pounds (27.3 kg) or more the bees will need that should be stored in these three boxes. If the honey stores are insufficient, feed, feed, feed sugar syrup to supplement their food stores.

The sugar syrup you feed now, however, is a different mix than you used when establishing your package in the spring. Mix a thick 1:1 syrup—one part (by weight or volume) sugar to one part water. Thick syrup has less water in it, and this time of year, the thicker the better. Thick syrup does not induce a build up, but rather storing behavior, as if the bees were simply moving already-made honey. Thin, spring syrup imitates nectar, stimulating a nectar flow and buildup and queen activity laying eggs.

If you are feeding to get those 60 pounds (27.2 kg) or more of honey stored, don't measure syrup; rather, measure the sugar used. Ten pounds of sugar in a gallon of syrup is only 10 pounds of sugar. Don't skimp here.

When you're finished feeding, the stored food should be at the sides of the broodnest and above the broodnest. And, the broodnest should be mostly in the bottom two boxes. Honey, then, should be in the outer two frames, plus some in the frames next to the outside frames. The top box should be almost all stored honey.

You can estimate how much food remains by looking to see where the bees are anytime you check during the winter. The closer to the top they are, the closer they are to running out of food. There's an old beekeeping axiom that says, "The top is the bottom, as far as food goes."

If your bees didn't read this, you may need to rearrange some honey frames while it's still warm enough to do so to ensure this configuration is close.

🐝 *A mouse that gets into your colony will cause all kinds of trouble. It will destroy comb (even plastic foundation), defecate in the colony, and raise young in the nest material it brings in. Be sure you don't put your mouse guard on too late in the fall and trap a mouse inside all winter.*

If you haven't already, replace the board that covers the screened bottom board. To improve ventilation, some beekeepers leave the back third or so of the screen exposed (by using a shorter board). This increases the opportunity for ventilation as the warm internal air draws fresh air up from the outside. If you choose to do this, however, you must enclose the space between the screen and the ground below so you don't get wind and blowing snow entering from below. Get mouse guards in place before mice move into your warm, dry, food-filled nesting spot. The expandable, metal types with holes rather than a slot are by far the best. The wood guards have slots, which are just not quite small enough to keep out a determined mouse.

By late fall, at least a full month before winter arrives according to the calendar, treatments should be complete and all assistance activities should be over. The last inspection makes sure enough food is available, and the cover is secured with a brick to foil strong winter winds.

If your colony doesn't have a good windbreak, you can build a temporary one to help. A stack of straw bales on the windward side is one way, as is a temporary fence of horticultural burlap and a few fence posts.

A word about ventilation—all during the winter, your bees continue to eat honey, perhaps raise some brood, and move around a bit. These activities normally generate heat (as does the very activity of clustering) and carbon dioxide and water vapor from respiration. This process occurs all summer, but the bees are constantly ventilating the colony for honey ripening, so it isn't an issue. In the winter, it can be an issue. Without a vent for carbon dioxide and warm, moist air to escape, it will naturally accumulate inside. The carbon dioxide builds up, displacing some fresh air. But the warm, moisture-laden air comes in contact with the unheated sides and inner cover and condenses—just like the fog that forms on a winter's day window when you exhale near it.

The water that collects there eventually runs or drips down back onto the bees. If you can, imagine a cold shower on a cold winter day in an unheated house. The bees struggle to stay dry, and keeping warm becomes even more problematic.

This situation can be avoided by what at first may seem a counterproductive technique. First, as you are preparing your colony for winter—checking stores, health, and the rest—turn the inner cover so the flat side is up. Then, between the inner cover and the edge of the top super, place a pencil or any block of material that raises the inner cover 3/8" (1 cm). Replace the inner cover and the cover over that when complete. Add a brick to hold the cover in place if winter winds tend to be blustery.

This pencil or small block actually serves two purposes. It allows the warm, moist air to rise and escape, leaving the colony dry. It also provides an entrance should snowdrifts cover the front door. Snow doesn't impede airflow into the colony but it will prevent bees from leaving to make cleansing flights.

If you haven't already, now is the time to tilt your colony forward just a bit by raising the back 1" (2.5 cm) or so. A 1-inch (2.5 cm)-thick board (or something sturdy, durable, and large enough to support the entire back of the bottom board) works well. The colony should have this forward slant so that melting snow or winter rains don't run into the colony and collect on the bottom board, adding to both ventilation problems and the difficulty the bees will have keeping a soggy bottom board clean.

This is the perfect time to check your records to make sure you haven't forgotten to do anything and, even more importantly, to note what you have done. A general checklist includes, but isn't limited to, the following items:

- Mouse guards in place
- Enough good food
- All medications removed
- Colony tilted
- Windbreak in place
- Inner cover propped up
- Outer cover secure with weight on top
- Feeders removed
- Screened bottom board cover replaced inside

When all the chores are done, all the equipment double-checked, the beeyard policed (any extra equipment removed), the weeds trimmed for the last time, and the bees as snug as they can be, it's time for you to take a break. You're finished for a while.

Cleansing Flights

In the colder regions, where snow cover is common, you will notice that on warm, sunny days (here warm may mean only 30°F [–1°C]) after a spell of colder days or snowy weather, bees will fly out for cleansing flights. Older bees, however, may become chilled and not be able to return to the colony; they soon fall and die in the snow. (Recall that bees require an air temperature of at least 50°F [10°C] to be able to fly for any distance.) On such a day, you may discover droppings and dead bees in the snow in front of the colony. This is normal and to be expected. The dead bees are generally the oldest bees in the colony and would have died shortly, even if they stayed inside. They often have a virus infestation—those bees made the ultimate sacrifice for the good of the colony.

Winter Checks

Depending on your winter weather, periodic checks may be routine or only rare. In the warmer areas, a quick check at least once a month for a couple of months is a good idea. You'll be briefly checking to see how the food is holding up, and generally, you'll be able to feed syrup nearly anytime. You can look at the broodnest briefly, especially on a 50°F (10°C) or warmer day to make sure a disease hasn't cropped up, that the queen is present and perhaps laying, and that there's enough food for the colony. You can feed syrup in the warmer areas if needed, or add frames of honey you stored from the summer for just this opportunity. Recall that when brood rearing begins, the amount of food needed can be incredible, and it's in early spring that most colonies starve because of the stress of brood rearing, so check food supplies carefully.

In the cold, snowy regions, winter inspections are not nearly as convenient. You just can't open a colony on a 20°F (–7°C) day. The cold air entering the colony, even for a few minutes, can kill brood, and the disruption can significantly disrupt the cluster. Moreover, no matter how cold, some bees will fly and they will be lost. But you do need to check. Pick the warmest, sunny day you can.

When you open a colony in the winter, all the propolis seals will be brittle and will crack loudly when broken. This definitely puts the colony on alert, so be sure to wear protective gear. Bees take a very unkind view of this disruption.

Open the top and puff some smoke in and reset the cover. Wait a couple of minutes. Then, tilt up the whole top box, pivoting it on the far edge of the box below, and look at where the bees are. If most are in the box below, you are probably in good shape. If most are in the top, pay attention . . . they are almost out of food and emergency feeding may be necessary. Look only for a moment and replace the box, puffing a tad of smoke to keep bees off the edges and getting squished.

Severe Winters

A hundred years ago, beekeepers routinely offered protection for their colonies from the hazards of winter. They put them in basements, buried them in trenches, wrapped them in straw, covered them with sawdust—in short, they insulated them from the ravages of winter.

Winter wrapping is a lost art. For some reason we've gotten away from those practices, though winters are not much different now than they were then, climate change notwithstanding.

There is a multitude of ways to offer winter protection for your colonies. Provide a wind break of evergreens, or a temporary break of landscape burlap, fencing, or straw bales. Wrap the colonies in roofing felt paper to provide wind and water protection, leaving the front door open for cleansing flights. Use the commercially available materials to wrap and protect your colonies, or create something similar. Long ago, beekeepers put a large box around each hive, and then filled the empty space with sawdust. You can do the same with home insulation.

A winter wrap will help the winter cluster so they can move to a new spot with food. Even a strong colony with lots of food cannot survive if they can't move to a place with lots of food (even if it's just the next frame).

Keep good ventilation in mind, a top and bottom entrance available, and ease of installing and removing. Put these protective materials on after the bees have formed cluster in the late fall, and leave them on until they can easily fly every day in the spring.

Winter Feeding

It has to be warm enough for the bees to break cluster in order to feed. If they can't move, they can't get to the food you provide. They'll need at least an hour, or two or three, of an internal temperature approaching 40°F (5°C). This isn't too uncommon on a sunny, still day, even when the outside temperature hovers around freezing. This is especially true if you have provided a winter wrap that is black, which absorbs the sun's warmth, warming the inside of the hive so the bees can move. The additional protection also slows the drop in temperature as the day closes, giving the bees a little more time to get settled for another cold spell, close to food and each other.

If the forecast is for continued cold, liquid syrup won't do much good. Even if they can get to it they'll need to remove the water, a nearly impossible task in the cold. At a time like this, they need solid sugar. You can put regular table sugar on

Fondant, made from table sugar and high-fructose corn syrup, makes an excellent emergency winter food. It can be sliced and put in food storage bags until needed. It has the consistency of medium-hard butter. When placing it on the colony, cut one side of the bag corner to corner in an X shape so that the bees can get to it. It is all sugar, so you know how much you fed, and the bees relish it.

the inner cover in an emergency, poured all around the inner cover hole. They may take this, they may ignore it, or they may even carry it outside as waste. It can be a gamble.

By far, the best food for this type of feeding is fondant, a mixture of sugar and high-fructose corn syrup, available from most bakeries. It comes in 40- and 50-pound (18.1 and 22.7 kg) boxes, wrapped in plastic. Remove the contents of the box, place it onto a cutting board, and roll back the plastic bag. With a large knife, slice off ½" or ¾" (1.3 or 2 cm)-thick slices, and put them in large, sealable plastic food storage bags and freeze until needed.

In late winter, if feeding is required, thaw the bags of fondant you'll need, and cut a corner-to-corner X on one side of the bag. Bring an empty super to your colony (and your protective gear and hive tool), and a thick slice of fondant in a cut bag. The slice will be nearly as large as the surface of the inner cover. Remove the cover and inner cover, peel back the cut flaps, and place the fondant cut side down on the top bars of the top super. Put on the extra super so the covers fit, add the inner cover and cover, weigh it down, and you're done. This material is a bit more costly than simple sugar syrup, but there is essentially no labor involved in mixing and stirring, and no pails to fill and refill. Feeding with fondant in the winter is easy, efficient, and effective.

You should know how much food (sugar) your bees need to finish the winter and feed all those young. Weigh the fondant slice and plan accordingly for more if needed. The fondant is all sugar, no water, so you'll know how much to feed.

If the fondant dries out, it will become rock hard, but setting the bag in the sun for a couple of hours will soften it up. Laying a sealed bag in a pan of warm water softens it more quickly.

Early Spring Inspections

Early spring may be as early as mid-January (or mid-June), or as late as late March (or late September) depending on where you live. If you can, time this first quick inspection at least a month or six weeks before nectar and pollen start coming in.

The bees are probably already raising brood, and the stress this puts on the colony has begun. Double-check for food, using sugar syrup or fondant as needed, depending on the weather. Pick the warmest day you can—around 40°F (4°C) or warmer—and open the colony to see where the bees are, whether brood is present, and how much food is available. Don't do a lengthy inspection, but do check on those three things.

As early as you can, monitor for *Varroa* with a sticky board for three days. If mite populations indicate a serious problem, which is unlikely, you will need several weeks for essential oil products to work, so plan ahead. Get a drone comb in the broodnest too, so it's there when the bees start producing drones. Put one in the super with the greatest number of bees and add more as the colony grows. Put in positions one and seven for either one super or two. You'll catch most of them that way.

If the weather's warm enough, you can remove all the boxes and clean the bottom board screen, removing the insert when finished checking for *Varroa* mites. It will probably have an accumulation of wax cappings, dead bees, and other material on it, all of which need cleaning off. Scrape it all out before you do your *Varroa* mite test. (Remember to collect what you scrape out, because if left near the hive, the scrapings will attract skunks and raccoons.) Remove the mouse guard and put the inner cover back without the ventilation spacer, flat side down, when you reassemble the colony.

If there are no bees in the bottom box, or even bottom two boxes, move the upper box with the bees down to your now clean bottom board, and place the two empty boxes above that. Bees will move up as they expand their nest, and this gives them room to expand. If you have a box of honey left, put that on the bottom, the bees in the middle, and the empty box on top. (With two boxes of bees, both full boxes will be on the bottom, the empty on top. Just make sure that the room they'll need is above them.)

If you need to rearrange frames, put honey on the sides, brood in the center, but, and this is important, leave empty frames in the middle of the brood so there's room to grow, all in the middle of two boxes. Visualize that the food should be close to the brood, all the brood and expansion space together in the middle, and empty space above. Pollen frames should stay next to the brood too. The bees will rearrange and straighten out what you messed up, but keep checking to make sure there is not a ring of honey across the top of the brood area below. That will slow the queen's laying because she'll run out of space, and crowding will occur.

Look Out for Swarming

A colony that swarms has the potential to greatly diminish your honey production because about half the bees in your colony will leave.

A healthy overwintered colony with frames of drawn comb, plenty of food, an expanding population, and a year-old queen is prime for swarming. Recall what motivates a colony to swarm: perceived limited space, lots of brood, lots of adult bees, an aging queen producing less queen pheromone, a honey flow just beginning, and cooperative weather. If you wish to control or prevent swarming, those are the factors you need to address.

One way to accomplish this is to requeen every year. Order your queens before the end of the previous year so you don't have to wait for a delivery date to open up. Have her arrive when the *Varroa* mite treatments are complete (if you needed to do that), when the antibiotic treatments are complete (if you did that), or almost (but not quite) as early as the queen producer can deliver. In fact, here's where patience is not only a virtue, but often a colony saving activity. Early spring in queen producing areas offers unpredictable weather, and with that comes unpredictable queen mating opportunities. Cold rainy weather reduces flight times, or cuts them off completely. Waiting for the third flush of queens from a supplier, rather than the first goes a long way in making sure you receive a well mated, healthy queen. Be patient.

Unpredictable weather, not enough time, and luck sometimes foul the best of intentions at swarm management, and the subtle steps toward swarming kick in before you can act.

The workers reduce queen feeding and foraging, egg laying slows and stops, but not before queen cells are started in the queen cell cups prepared just for this occasion.

You'll see queen cells hanging from the bottom of brood frames, little or no open brood, and little activity. Once the queen cells are capped (nine days from egg to sealed cell), the door opens and the swarm takes off.

If all the queen cells are still uncapped (or all those you can find), you can make a split of the colony and, maybe, thwart the swarm. Once queen cells are well started, the colony is probably going to swarm. Watch for it, and perhaps you can retrieve it.

There are as many techniques to control swarms as there are beekeepers. Basically, they focus on separating the brood from the bees, and they all entail a fair amount of work. The best choice for managing swarms is to prevent the behavior before it begins, or catch the swarm after it leaves.

Making an Early Spring Split

If expansion or swarm prevention is in your plans, early spring is the time to consider this activity. If your colony has at least four frames with open brood and four with sealed brood, you can divide the colony into two, much like dividing a daylily or other perennial.

Have ready a screened bottom board, four medium supers (two with frames), an inner cover and cover, and a feeder for the split colony. Order a queen to arrive close to the time you want to do this. Here early queens are necessary, and can be a risk. Watch them carefully.

When the queen arrives in the mail, or when you pick her up from your local supplier, assemble all the equipment near the colony you want to divide. Shoot for late in the day so as many bees as possible are home. Place the bottom board on the hive stand and an empty super (keep the frames close) on top of that.

Smoke your existing colony—now called the donor colony—remove the cover, inner cover, and top box, and locate the box below with the most brood.

From the donor colony, carefully remove two frames (three are okay, too) of open brood (there may be some sealed brood in the center), complete with the bees that are on the frames. Make sure the queen isn't on any of the frames you move—if you find her, let her run back into the box. Put those frames in the center of the split box. On either side, place frames with sealed brood and the bees from the donor colony. This should give you four or five frames of brood and some bees. Take another frame of open brood, hold it over your new box, and give it a good downward shake so that most of the bees fall off and into the box. You may need to do this two or three times to get enough bees in the box.

Add one frame of honey, and one of honey and pollen on either side of the sealed brood for convenient feeding.

When complete, replace frames on the edge in the bottom box of the new colony, put another box with frames (drawn foundation are better, with three or four borrowed from the top box of the other colony), install the new queen using the techniques explained earlier, put on the inner cover, the feeder (filled with 2:1, water: sugar), and close it up. Place a wooden entrance reducer in the split box entrance hole so

only the smallest opening is available, and stuff some grass into that hole. And leave it closed for at least 48 hours. Best, if possible, is to do this as late as possible on day one, and let the bees open it up. But try to keep them in at least two days, and if the weather is iffy, three is just fine.

Close up the other colony, first pushing together the remaining brood frames and replacing those missing on the edges. You now have two colonies where there was only one before. This is also a good time to requeen the donor colony. Ordering two queens isn't a bad idea. And, if one doesn't work out, the two colonies can be rejoined as a single unit.

Examine your new colony on the same schedule as you did your new package last season, checking the queen(s) (she, or they, should be accepted with little difficulty), food, and the feeder. After a day or two, the bees will have cleared away the grass and will begin foraging as if this is exactly what they expected.

After the right amount of time, make sure the new queen is released and begins laying. Once that's underway, treat your new colony like your package and enjoy watching the growth.

No matter what you do, though, sometimes that colony will swarm, either before you get ready to do your manipulations, or even after you've done everything right. It happens, but you may be able to retrieve the bees that left, that is, if you're home, if you know they swarmed (check again for the signs of swarming), if you find them after they swarmed, and if they're not 50' (15.2 m) off the ground. Simply, they may leave and you're not aware of it until later, and you have contributed to the diversity of insect life in your community.

If you happen to catch that swarm or another, and if expansion plans are not in your future, you have some options. Give the bees to someone who is planning on expanding, start a new colony to sell, or join them to one of yours.

How to Catch a Swarm

When a swarm leaves a colony, it usually doesn't travel far before it stops to rest. The swarm will gather in or on almost anything: a tree limb, fence post, family picnic table, car fender, lamppost, or street sign.

Scout bees can be seen dancing on the surface of the swarm, indicating the location of a place they have found. The more vigorous the dance—like the dance pointing to a nectar source—the better the location. There may be several dancers performing at once, indicating several options to be weighed. The bees seldom make the final decision immediately. Sometimes, it takes two or three days but a few hours or overnight is more frequent.

Once people know that you are a beekeeper, you will be sought out as the person to deal with a large mass of honey bees hanging someplace. In fact, if increasing your stock is

a goal, retrieving swarms is a viable and inexpensive method to do so. Knowing when swarming begins in your area helps so you can plan on having your equipment ready. Beekeepers often notify local fire departments, police departments, and county extension offices that they will take swarm calls.

You will need some basic equipment before going on your first swarm call. First, you'll need a box in which to put the bees. An unused super works well most of the time, but you'll need a bottom and top for it. For adequate ventilation, use a top that's screened rather than solid; using a window screen wedged in the front door works for short drives home in a truck. You'll want a far more secure box and closure if you are putting the box in the backseat of your car. The super should have only three or four frames in it, no more.

You'll also need your smoker, bee suit, and gloves, a full spray bottle of sugar syrup, and perhaps a piece of tarp or carpeting. You should have a means of holding all the pieces of the box together snugly, such as a ratchet strap or a bungee cord, for the trip home. Take pruning shears, to remove stray branches, or to cut the entire branch on which the swarm is sitting is almost always handy. Have a bee brush along, too.

Swarms that are very high in trees are seldom worth the risk of wobbly ladders and precarious positions. Safety is paramount here. A 6' (1.8 m) stepladder is as high as you should go if you're not used to this activity. Even then, you should go only halfway up.

Position the box (with the frames removed) under the swarm, as close to the bottom of the cluster of bees as

Swarms leave the nest and fly a short distance to get organized, making sure the queen is with them, and evaluating the places scouts have found that they may soon call home. Often they land high—the tops of trees, the sides of buildings, church steeples— but they can land just about anywhere imaginable (car bumpers, bushes, beneath picnic table benches). If retrieving your bees from great heights is an issue, let the bees go. It is not worth the risk. There are swarm catchers made from long poles and large buckets with a lid you can close from the ground. When a swarm lands close to the ground, on a small shrub or bush, gathering them can be easy, or very, very difficult. Either way, the thrill of gathering several pounds of free bees, and being a hero for those who believe they have been saved by the beekeeper, goes a long way.

Swarm Call Checklist

If you're on a swarm call list for the fire or police department, you need to be prepared.

Here are some common questions to ask the caller before you leave:

1. Name, exact address, and cell phone number of where the bees are, or who the caller is
2. Are they honey bees, and how long have they been there?
3. Where are the bees, exactly—how high, on what?
4. Are they posing a problem—children, traffic?
5. Who might own the bees—a nearby beekeeper?
6. How big is the bunch of bees? Softball, basketball, and beach ball are sizes most everyone knows.

You'll learn more questions if you do this often. And always have a paper and pen ready.

Things to tell the caller:

1. Keep a safe distance from the bees or stay indoors.
2. Do not spray them with water or pesticides.
3. Have somebody meet you at the location. Tell them what kind of car or truck you have. They may be inside.
4. Tell them the bees will leave soon because they are just resting.

What Is Your Liability?

A sobering thought on capturing swarms is liability. What happens if you are injured capturing a swarm on someone else's property? What happens if someone else is injured while you are removing the swarm? The likelihood of either of these events occurring is rare, but they do exist and they often end up in court. This must be considered when retrieving a swarm in a busy neighborhood, on public property, or from property owned by a business.

Another consideration is that you should never, ever charge for the opportunity of removing a swarm unless you are in the pest control business. Moreover, never offer to pay for a swarm if you remove it. These acts can be interpreted as a business venture and can give you an entirely different liability exposure.

People and honey bees mix in odd ways, and it can be difficult to predict how a situation will develop once you arrive at the scene. Don't be afraid to walk away from something you consider dangerous or from someone who is pressuring you.

possible. Use the top or shelf of the ladder to hold the box if the bees are that high. Spray the swarm with sugar water several times to keep them from flying around. This also feeds them if they have been in that location for more than a couple days, and slows down any defensive activity. Place a piece of carpet underneath the box. It helps so they don't become lost in the grass or other weedy growth.

Ideally, you can lower the swarm or raise the box so it actually has some of the swarm in it. Once everything is secure; you, the ladder, the box, any bystanders, and the swarm and box are as close as possible; and any stray, small branches are removed, you're ready. What you're going to do is jerk the branch on which the swarm is positioned so they are shaken loose and the whole of the clump falls into the box. Somewhere in the clump, with luck, is the queen. Once the bees realize she is present in the box, they will stay with her.

When the branch is jerked most of the bees will fall, but some—perhaps many—go airborne. They'll return to the spot on the branch to regroup. If the queen is there, they'll stay. If not, they'll look for her. Keep the box as close to the original location as possible. Some bees will miss landing in the box.

Once the mass of bees is inside the box with the queen, you'll notice bees at the front door fanning the here's home pheromone, attracting those bees still in the air or lost.

In short order, almost all the bees will be inside, especially if it's early evening (a good time to do this, if you can pick the time). Later, you can close the front door, secure the boxes, and head for home. Swarm calls are, however, seldom ideal.

Obtaining swarms, whether your own (yes, yours will one day swarm) or from some other colony, is seldom predictable. Swarms will cover a fence post, become entangled in low shrubbery, or cling to the side of a house. All of these locations require that you brush or push the bees into a container. This puts lots of bees in the air again, in spite of the sugar syrup. And, while pushing, if one or more get crushed in the mess, there's alarm pheromone added to the mix.

Bees On a Vertical Surface

A swarm may land on the wall of a house, a fence, or even the side of a car. If this is the case follow this simple procedure. If you can position your box directly under the bees do so, but often you can't. Obtain a piece of cardboard or plastic and position the box as close to the wall as possible. (Actually, add this to your swarm kit. Make it about 1 foot [0.3 m] wide and 3 feet [1 m] long, foldable in the middle.) Place the sheet of cardboard on the edge of the box, or even resting on the bottom of the box if the bees are close, and lean it against the wall, beneath the bees, or even against some of them if the piece of cardboard is too tall. You can fold it if they are really close. What you are going to do is, first, spray the bees with sugar syrup until almost too wet so they can't fly and are busy cleaning. Then, using your brush (you brought your bee brush, right?), begin sweeping them slowly down, beginning as close to the bottom of the clump of bees as possible. The bees will fall, sliding down the cardboard directly into the box. Go slow. Very slow. Continue sweeping bees, raising the cardboard as you move up the wall. Spray the bees again, both on the wall and in the box to keep them from moving. Continue until you have all the bees off the wall. Position the box so the opening faces the wall where most of the bees are that fell and missed the box. Replace the cover, wait awhile— that may be twenty minutes, or an hour—and pretty much all of the bees will be inside. Close the entrance and head for home.

If swarm calls are in your future, begin by helping someone experienced, if possible. And you should always be prepared for the unpredictable.

If you extract your honey, you'll have both honey and wax when you're done. (If someone does it for you, his or her payment may actually be in wax or honey.) The wax is what is removed when you cut off the cappings from the honey-filled cells before you put the frames in the extractor. These cappings are usually collected in an uncapping tub that lets much of the honey drain off, separating the cappings wax from the adhering honey. (Now you see the wisdom of lining your uncapping tank with a mesh lining: It keeps the wax and lets the honey go.)

You have the choice of separating the remaining honey from the wax, or simply discarding both. That may not be the best choice; there is a lot of honey clinging to the wax cappings, and the best beeswax is found in those wax cappings. If you choose to keep the wax, there are several ways to proceed.

The simplest method is to gather the corners of the mesh filter that lines the uncapping tank and shape it into a large bag. Tie the "bag" closed and suspend it over a clean bucket to let the honey drain. Let it drain in a bee-free area in a warm location for a few days. Add the honey to your crop and clean the wax in water. Once the wax is clean, freeze it to destroy any wax moths. You can then keep it for later, give it away, or melt it down yourself. Whatever you do, do it fairly soon so the wax moths, or, if you live in the southern parts of the United States, Australia, or South Africa, small hive beetles cannot cause problems. (This wax can coat next year's plastic foundation, as well.)

One way to remove cappings is to use a warm knife to slice them off. The beeswax cappings fall into the cappings tank. The tank has a platform in the center that holds the wax above and lets much of the honey drain below where it can be removed without wax. The wax is then washed in water and then melted to be used for many purposes.

Terms of Uncapping

The terms *decapping* and *uncapping* mean the same thing: mechanically removing the beeswax coverings from the honeyfilled cells on a frame. The coverings are called *cappings*. When melted, the beeswax is called *cappings wax*.

Be Safe

To melt the wax there's one golden rule. Every beekeeper knows this rule, and many don't follow it: NEVER melt beeswax over an open flame. Period. If the temperature of the melting beeswax rises past its melting point and boils over the sides of the heating container, that liquid wax, when it makes contact with the flame, becomes a torch, burning uncontrollably. All the wax will soon catch fire, and that burning wax will soon spread over your work surface, setting fire to anything flammable it comes in contact with. Also, wax vapors from overheated wax can explode if exposed to an open flame. (Boom! One less beekeeper.)

Beeswax can be melted safely over a double boiler. For the amount of wax you will generate with a few colonies you won't need a huge rendering facility. (*Rendering* is the term that encompasses all aspects of transforming beeswax from the hive into a clean block of wax.)

Find a Safe Place

If you can, melt wax in the driveway or another outside space where a few wax spills won't matter, and where the heat can dissipate.

The smell of melting wax will attract your bees, so keep that in mind.

If you can't set up outside, choose a spot in the garage or the basement. Never, ever do this in the kitchen. Spilled wax on the floor, stove, sink and sink drain, shoes, counter, or places you never imagined will give you grief for weeks.

If there's no place to do this safely, drain the honey, wash the wax in a pail of water, drain it, bundle in a plastic bag, and put it in the freezer. Wait until a safe place is available or have somebody with experience do this for you, rather than do it dangerously.

In your safe-melting area, set up a workbench in your space. Saw horses and a few planks, covered with newspapers or other disposable covering, work well. I use a leftover piece of wood siding from a garage update cut to about 4' × 2.5' (1.2 × 0.8 m). One end sits on a sawhorse while the other is propped up on unused bee boxes. It's not pretty but it doesn't need covering, and it works just fine as a temporary workbench for many tasks.

Types of Wax

Before you begin melting raw, uncleaned wax, sort it by color. Burr and brace comb creates nearly white wax. Cappings wax is lemon colored. You may have darker wax from brood frames. Mixing colors creates dark wax. The darkness comes mostly from propolis and fine dirt particles in the wax that won't filter out. Propolis also makes candles burn a bit erratically, can give lip balms an off flavor, and could make creams and lotions a darker tinge that you may not want. Dark wax is good for making soaps because of the colors, for household uses such as lubricating drawers and the like, and for polishes and water proofing lotions. Save it for those uses.

Only melt darker wax in a solar melter. A solar wax melter removes wax from plastic foundation and separates the wax from the cocoons, wire, and propolis on brood combs. Also, the uncontrolled heat inside a wax melter may lead to darkening the wax, which is what happens when wax is heated to extreme, 200°F (93°C) and higher.

Try to melt beeswax as few times as possible.

Wax from seven beekeepers will be seven different colors.

Melting Beeswax

Use an electric heat source—not gas, not flame. If you are outside or away from a dedicated outlet use a heavy-duty grounded extension cord. I use a double burner hot plate. For more detail, see the section on melting wax for coating plastic frames—it's the same setup. I can melt a lot of wax when I'm in my industrial mode. Containers should be made of aluminum, enameled steel, or stainless steel. Don't use copper, iron, or nickel as they may impart a dark color to the wax. If in doubt, first test with a small amount of wax to see if the color changes.

If rough wax (from the wax melter, chunks of comb, old candles, and so forth.) is being melted for the first time, I filter it as I'm pouring it from its melting container. Extraneous material, such as dead bees, cocoons, wood, and various unrecognizable objects can be removed with a coarse sieve. The next round of filtering can be through old sweatshirt material (fuzzy side up), milk filters, or even a few layers of paper towels. (Wax-soaked paper towels can be used later to light a smoker, after melting off most of the wax.) Tulle fabric also works well, as do a variety of other fabrics. The finer the mesh, the slower you have to pour. Increasing the surface area of the filter by using a larger catch pan speeds up the process, as does using a filter that concaves itself almost to the bottom of the receiving container. For small lots I use a smaller water pan and a single burner.

When the bench is set up, the electricity source safely secured and grounded, the hot plate positioned, and the water pan(s) about half full of water, place your wax in the containers that are sitting in the water and turn on the heat.

Keep the heat on until the wax melts and becomes clear. (This may take a while.) Have your receiving container ready with the filter in place, either supported or held in place with paper clamps or rubber bands around the container. Make sure the filter is secure—the filter will become heavy with cooling wax.

When melting rough wax the first time I heat it until it is clear and then pour it through a double layer of paper towels directly into a pan that I can reheat it in again. When finished, I put this second pan back in the hot water and heat it until clear. Then again pour it through the finest filter I have into a storage container. This way, there is no debris in the wax, and I can accomplish it all in one sitting.

You can heat wax in a pan with some water in it already. That pan sits directly in the other pan of water being directly heated. Wax is lighter than water and will float to the surface of the pan when it melts. When all the wax is melted ladle it out into your receiving container. Most of the debris will settle out to the bottom of the container, and you can scrape off any debris that was not removed.

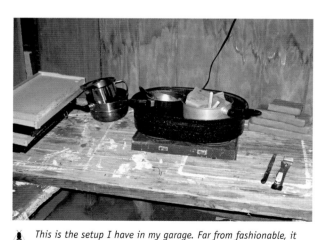

🐝 *This is the setup I have in my garage. Far from fashionable, it is extremely functional. From left to right: frames to be waxed, a pitcher and a pan for holding clean, melted wax, the turkey roaster half full of water, a small aluminum sieve, a large sauce-pan for melting chunks of wax. (This is essentially a double-boiler.) Surrounding it are blocks of clean wax to be melted, knives, and a hive tool for breaking the larger blocks of wax to fit the pan.*

Waxing Plastic Foundation

Place the frame close to the hot plate on the work surface, raising one end. Dip the sponge brush in the wax and squeeze the air out. Lift the sponge, tap a few times to release excess wax, and quickly bring the brush to the foundation in a sliding motion, moving across the surface immediately. Most of the cell edges have wax without the cell having wax in the bottom. Move fast enough so the wax doesn't puddle in the bottom, but not so fast you don't leave any wax on the edges.

First, to remove debris from the melted wax, pour through a sieve.

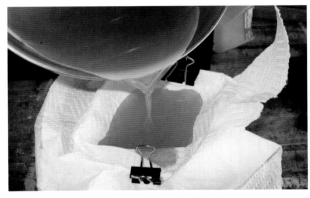

Melted wax is being poured through a filter to remove any remaining debris.

A pan of melted wax sitting in a water bath: Note the ring of wax cooling around the edge of the pan. This wax is destined to be brushed on a sheet of plastic foundation.

Brush the foundation in a sliding motion, moving across the surface immediately.

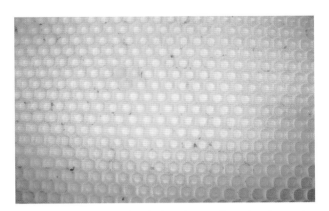

A well-waxed plastic foundation—beeswax is on the edges and not the bottom of the cells. Bees will make good comb with this.

About Solar Wax Melters

You can make a wax melter or purchase one as a kit from suppliers. Plans abound on the Internet and they are fairly easy to build. Paint the outside and inside black (some say paint the inside white, but I've found black works better when using a thick plastic covering), make sure the cover is tight or you will have bees in it all the time, and place it so it gets as much light as possible all day. Cover the top with corrugated plastic rather than glass, and make the metal tray that the frames and wax chunks sit on as large as possible. Put in a large catch pan and have two or three—this may be the pan you put in your wax melting pot on the hot plate so make sure it's aluminum or stainless. If you get serious about this tool, make it big (I've seen homemade boxes that use a regular storm door with windows for the top), make the pan slope between 30 and 35 degrees so the melted wax moves away from the frames and other refuse fairly fast. If you can, put a screen on the opening at the bottom of the tray to catch at least some of the bigger chunks of impurities that are in the wax—dead bees, chunks of old comb, pieces of wood will all be screened out.

Warning: Those all-plastic frame/foundation units will be ruined in a wax melter. They will twist and warp and bend and nothing will get them back to their original shape. Some styles of wooden frames with plastic foundation will twist and bow, others won't at all. Be careful with your frames with plastic foundation until you know how they will react to the heat.

🐝 A solar wax melter is a tight box covered with translucent plastic that and painted black (left). Inside temperatures can easily exceed the 150°F (66°C) temperature needed to melt wax. Frames or wax are placed on the slanted tray, and melted wax is collected in a small pan (above). This is a great way to collect raw wax for candles, lotions, frame coatings, and more.

🐝 A typical solar wax melter is simply an enclosed box that holds wax on a slanted tray so that when the internal temperature warms to over 150°F (66°C) the wax melts and drains off the tray into a catch pan at the bottom, leaving any debris on the tray. Wax melter pans are messy and an attractive nuisance to bees because of the smell of honey and wax.

🐝 Wax straight out of the wax melter needs to be remelted and filtered to remove all of the extraneous material so it is clean, burns well in candles, and is suitable for lotions.

Dealing with Cappings Wax

When the wax has melted it's ready to pour—but what do you pour it into? Storing wax in your melting pans isn't a practical option. Suppliers sell special wax pans. After these pans are partially filled with hot wax and the wax cools, the solid block of wax slips right out. Empty, clean, paper or plastic milk containers work well, as does any container that can withstand the heat of melted wax. If in doubt, test the container with a small amount of melted wax first.

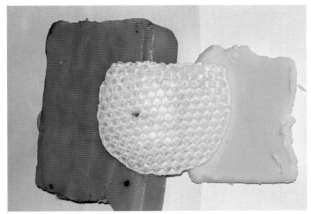

Avoid mixing waxes when melting—you'll end up losing any distinctive color. Left to right: dark wax from old comb; fresh new comb removed from a swarm box; and lemony cappings wax.

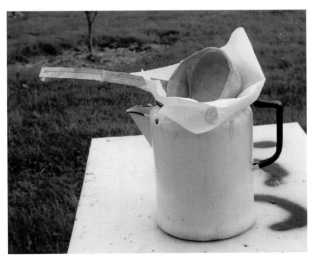

Paper milk cartons work well for holding beeswax until it cools and is ready to be made into candles, creams, or soap.

An old enamel coffee pot is perfect to catch wax in if you are going to pour it again. This one has a paper towel filter. Note the color of the wax. This soft lemon yellow is the highest quality, most sought-after color for beeswax. This is what melted cappings will look like.

Properties of Beeswax

- Beeswax melts at about 145°F (62.8°C). This temperature will vary a bit depending on air temperature and amount of debris in the wax.
- Density is about 0.96, whereas water is 1.0, so wax will float on water.
- Cappings wax, when cool, will be a soft lemon yellow color. Wax from old frames and bits of burr comb will be darker and will contain melted-in materials such as propolis. Do not mix cappings wax and old wax.
- Wipe up fresh spills with paper towels.
- Scrape up cooled drips and dribbles with a sharp-edged tool, such as a single-edged razor blade.
- To remove small spills and the thin film that remains after wiping and scraping, use a petroleum solvent specially made for wax removal available at most stores that sell candles. A final rinse with hot, soapy water will finish the job.
- When you mix softened tap water with melted beeswax, the wax and water react, resulting in mushy wax that is unfit for anything. To avoid this problem, don't use your tap water. Bottled water or even rainwater is preferred. If you are melting a lot of wax and using a lot of water, you'll want to explore another melting method—the solar wax melter or double-boiler.

Making Candles

Candles made from pure beeswax are unsurpassed in fragrance; they provide long, clean, smoke-free burning; and simple beauty. Numerous books and reference materials on making all types of candles are available. If this is a part of beekeeping that you enjoy, you'll want to explore more than the simple candles shown here. But these fundamentals for making candles provide a great place to start.

Before you begin, you'll need some supplies, notably the molds in which you will make your candles and the wick that is the heart of every candle. Many companies sell all you'll need, and you can find some in the beekeeping journals, on the Internet, and in candle books.

The wicks you'll need come in various sizes, shapes, and materials. The primary consideration is the diameter of the wick. If it is too large, the candle will burn with too large a flame and smoke; if it's too small, it won't burn at all. And to confuse matters, different catalogs use different measuring systems.

Most will give the size (diameter) of candle a particular wick works best with, but some don't. Get several catalogs for comparison before you begin.

Mold designs are infinite, so finding what you want—from elegant tapered dinner candles to a candle shaped like an ear of corn—is fairly simple. The most popular and most usable molds are made of polyurethane. These are split up one side so the finished candle is easy to remove. They are held snugly together during pouring with rubber bands so that wax doesn't leak out, and any seam that remains can be easily removed. Some shallow molds don't need the slit—you can simply push out the finished candle. These easy-to-use and fairly long-lived molds are great for beginners.

When making candles, lotions, potions, creams, and balms, use the cleanest wax you have. Twice-filtered is my recommendation. It must be clean and free of debris, pollen, and more.

Use the double-boiler techniques previously described (page 154) to melt the wax, for remelting or the second cleaning. If for a remelt you may want to use something made for pouring, maybe an old enameled or aluminum coffee pot. If for the second cleaning, melt in one of the original pots, and pour into the coffee pot or other container. Pour it through a filter again for the second cleaning. This wax is then ready to be made into candles, lotions, or other products.

A huge selection of molds is available for making candles for every taste and occasion.

While the wax is warming, prepare your mold and wick. Most polyurethane molds don't need a spray of mold release—you'll learn if yours does or not. (Check the instructions that came with your mold.) String the wick up through the bottom of the mold. You may need to push it through with a nail, wire, or hooked needle made especially for this purpose. Pull up the wick and attach it to a support. A large bobby pin works well. Secure the wick in the very center and pull it taut. Wax probably won't leak, but protect the surface the first time if you're not sure. The opening of the mold will be the unseen bottom of the candle, so the finish isn't critical, but it can be easily repaired if it turns out uneven or unsightly.

When the cleaned wax is just melted, it's cool enough to pour. First place the mold on a small tray or other covering. Slowly fill the mold, avoiding splashes and bubbles in the wax. The slower the wax cools, the less likely the candle is to be deformed or to crack. If, after the wax has cooled, the bottom recedes, gently refill the mold and cool again.

When completely cool—about a day at average room temperature—untie the wick and carefully remove the candle. Don't be in a hurry, especially if you are using an intricate mold with lots of detail. Remove any rubber bands used to hold the mold together, and gently pull back the sides of the mold, exposing the cooled candle.

Pour the melted wax through the final filter, shown here secured with clothespins. Protect your surface, and place the container with the clean wax on a hot plate switched to the warm setting, if needed.

Choose the correct size wick. This may take some experimentation, but it's all part of the learning process.

🐝 *When the mold and wick are ready, fill the mold with barely melted wax. Top off the mold with more melted wax if the bottom shrinks after cooling. After the wax is completely cooled, carefully remove the candle from the mold.*

🐝 *You can easily make seasonal ornaments with embedded hangers. Use decorative string or ribbon for the hanger. Fill a decorative candy mold about halfway, lay the hanger in the wax, and finish filling. The loose ends of the hanger are hidden and secure.*

Lift up the candle and pull enough wick into the mold to make another candle—this trick eliminates the wick-threading task next time.

If the bottom of the candle is uneven, you can smooth it out. Warm a small pan with aluminum foil on the bottom on the hot plate and rub the bottom of the candle until smooth and even. Pour the melted wax back into your container. Your candle is finished and ready to burn.

Before you start sharing these wonderful gifts with friends and family, or even selling them, make sure they work. You'll have to test-burn a couple of each kind you make to make sure the wax is clean and you have the right-size wick. You don't need to burn them completely if wax is in short supply, but use at least half when testing a small candle.

If the candle burns with a steady, clean flame, and without smoke, drips, or sputtering, you can be fairly certain you've chosen all the right ingredients. But if it doesn't burn well, figure out why: it's either dirty wax, a poorly size wick, or air bubbles in the wax. Make adjustments, such as cleaning the wax again or choosing a larger or smaller wick. Melt the test candle to use again.

Candle Safety

- Never leave a burning candle unattended.
- Place candles out of reach of children and pets.
- Discard (or remelt) a candle stump when it reaches about 1" (2.5 cm) in length.

Making Cosmetics

Beeswax is an essential ingredient in many cosmetics. It is used because it is an all-natural, hypoallergenic substance that adds body and texture to the finished product, has a delicate aroma, and is a wonderful skin softener and protector. It is used in hand, foot, and body creams; lipstick; eye shadow; lip ointment; and fine soaps.

Most of these creams are easy to make, even in the kitchen, and are a useful way to use small amounts of beeswax. The other ingredients are available at health stores, as well as many sources available on the Internet.

Here are a few simple recipes that use your beeswax and honey for homemade beauty products. The Internet is overflowing with ingredient and container sources and you can find anything you want in any quantity, price range, and taste. You supply the best raw ingredients—your honey and your beeswax—and the rest is easy.

A creative word of advice: Whether enjoying these recipes or others, keep good records. First and second time you make them, follow the recipes exactly to see what the finished product comes out like. Then, experiment with the ingredients, varying the quantities, the fragrances, and the texture additives. Keep track of how each modification affects the final product. Too often a new, experimental recipe is made, by accident or design, and the result is fantastic, but you can't remember how to replicate it.

Write down what you did, exactly, every time.

ELLEN'S HAND BUTTER

This simple recipe is from Ellen Harnish, half of the BeeOlogy Company cosmetic and soap team. Ellen is the marketing arm, producing labels, packaging, and finding outlets. She also makes lotions and creams and is an encaustic painter, using the beeswax and honey produced by the production half of BeeOlogy, Dave Duncan (seen catching that swarm on page 150) is the beekeeper, candle-making, and soap-making part of the business. This makes a rich, hand butter for nighttime use. Apply just before bed-time for softer hands in the morning.

Ingredients

½ cup (112 g) cocoa butter

½ cup (112 g) beeswax (Use cappings wax, or the lightest color wax you have.)

4 tablespoons (60 ml) avocado oil

½ teaspoon fragrance or essential oil

Melt all the ingredients except the fragrance in a 2-quart (2 L) stainless steel pan. Stir constantly until the beeswax (the last to melt) is melted and completely blended with the other ingredients. Do not heat to more than 120°F (49°C); use a candy thermometer to monitor. Add the fragrance (this so it doesn't volatilize before pouring and you lose the aroma). Pour into decorative containers and allow to cool.

JEANNE'S HAND CREAM

Jeanne Schell keeps bees in Medina County, Ohio, with the help of her daughter, Katie. Jeanne says the beauty of her simple hand cream recipe is that you can substitute varying amounts of shea butter or cocoa butter for the palm oil and coconut oil if you want to customize the aroma and thickness of your cream. Experiment with varying amounts of oils and butters to get just the right mix for your hands. Just make sure that the combined amount of oils and butters, other than base oil (olive oil), equals one full cup (235 ml). You can also vary the amounts of the ingredients to make more or less of this cream, but the basic recipe fills about one dozen 2-ounce (55 g) containers—more than enough for your use, with lots of extras to share.

Ingredients

2 cups (475 ml) olive oil (the base oil)

¼ cup (60 ml) palm oil

¾ cup (175 ml) coconut oil

6 ounces (170 g) beeswax

2 vitamin E tablets or 6 drops of vitamin E oil

40–50 drops essential oil(s) (optional)

This easy-to-make hand cream for everyday use will protect and soothe your hands after the rigors of home and garden tasks.

Combine the oils in a 2-quart (2 L) stainless steel sauce pan, and stir over medium heat until the oils (and butters, if using) are melted. Add the beeswax to the pan, and stir until melted. Then test the mix by dropping five or six drops onto a sheet of waxed paper, let cool, and test the hardness. If it is too hard, it will be difficult to rub onto your skin, and your mix will need a bit more base oil added to soften it. If it is too soft or greasy, add a bit more beeswax to stiffen it.

When the hand cream has the exact thickness you want, remove the pan from the burner and allow it to cool until the cream begins to harden on the sides of the pan. Then stir in two vitamin E tablets or six drops of vitamin E oil to enhance the healing properties of the hand cream.

For a fragrant hand cream, add any combination of essential oils after removing the pan from heat, or simply rely on the subtle fragrance of beeswax and the oils in the basic recipe. A popular blend that offers a whiff of fragrance without being overpowering contains 15 drops of lavender oil, 15 drops of rosemary oil, and 15 drops of geranium oil. You can use other oils, or vary the proportions of these, to suit your taste.

When well mixed, pour the cooling cream into individual 2-ounce (55 g) containers and allow it to cool, uncovered, overnight. Then cover the containers, label them, and store at room temperature, out of direct sunlight.

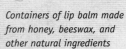

Containers of lip balm made from honey, beeswax, and other natural ingredients offer a fragrant cure for chapped lips.

Nancy's Lip Balm

Nancy Riopelle makes and sells lip balm from the beeswax and honey that her husband Buzz (yes, that's his name) produces from his 100 hives each season. She sells lots of this recipe at their local fairs and farm markets in Medina County, Ohio, each summer and fall. This recipe makes enough lip balm to fill 100 lip-balm tubes (0.15 ounce, or 4 g each), or 65 pots (0.34 ounce, or 9 g each).

Ingredients

1 cup (225 g) shredded beeswax

14 ounces (425 ml) coconut oil

5 tablespoons (100 g) honey

5 tablespoons (70 ml) pure vanilla extract

Heat the wax in a saucepan over low heat to 150°F (66°C). In a separate saucepan, heat the oil to the same temperature. When both are heated to the proper temperature, add the coconut oil to the beeswax, remove the pan from heat, and stir steadily until well blended. Then add the honey and the vanilla extract, and continue to stir until well blended. Pour into tubes or tubs, allow to cool overnight, and then cap the containers and store at room temperature, out of direct sunlight.

Jeanne's Famous Favorite Foot Cream

Jeanne and her daughter Katie make creams and soaps using beeswax and other products. They've been at it for some time and have simplified the process quite a bit. They make and sell their products at craft fairs, farmer's markets, and other places. This is Jeanne's favorite foot cream recipe.

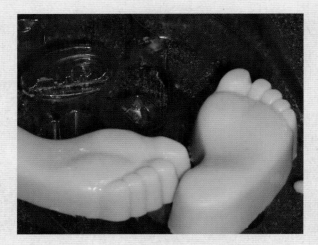

Ingredients

Scale that weighs in ounces or grams

Stainless steel 2-quart (2 L) pot

Plastic stirring spoon

Molds

Mold release

2 ounces (55 g) beeswax (small broken pieces)

2 ounces (55 g) deodorized cocoa butter

4 ounces (113 g) shea butter

Peppermint essential oil, 20–30 drops (lighter to stronger aroma)

Shea butter is a pale, solid fat pressed from the seeds of the shea tree, commonly used in foods, creams, soap, and even candles. The unrefined product is normally used in creams and has a hint of chocolate aroma. Add to that the chocolate aroma of cocoa butter, the fragrance of beeswax, and a hint of peppermint, and you have a very pleasing mix.

Melt cleaned beeswax and 2 ounces (55 g) of cocoa butter in the stainless steel pan until they are just barely melted (1). Use only low to moderate heat. When ready for the next step, the mix will be just barely clear.

Add 4 ounces (113 g) of shea butter and stir, allowing it to melt in the beeswax–cocoa butter mix. Reduce the burner temperature and let the mix slowly cool until it reaches the can-just-barely-pour stage—a thermometer isn't needed here.

At that stage, add the peppermint oil. You want the mix as cool as possible so it doesn't volatilize the fragrance of the peppermint oil (2).

Spray your molds with mold release while the mix is cooling, and set the molds on a protected flat, solid surface. A kitchen counter is ideal, but this takes some space while cooling. If you need to move them when finished, place them on a board beforehand so they don't tip or squeeze. When ready, fill the molds full (3). Let cool for twenty-four hours (4). If the bottom is uneven, level using the hot pan technique for candles, but not so hot it melts too fast. When cool, remove and place in a package that seals the bar—for home, a sealable plastic bag; for sales, a clean plastic wrap, plus ribbons and bows—so the fragrance doesn't escape too fast (5, 6).

Use Jeanne's foot cream for rough and dry spots on hands, feet, elbows, and knees. It is fragrant and pleasant to use—and smell.

When all done, the pan and molds can be washed up in hot dishwater without fear of clogging drains or ruining pots or spoons. Spills, too, clean up well with soap and water. From start to finish this project, from gathering your materials to setting aside the filled molds, takes less than a half hour. Beeswax creams are fast and easy to make, useful for friends and family, and even profitable.

(1) Weigh out 2 ounces (55 g) of beeswax, and 2 ounces (55 g) of cocoa butter. Put them in a stainless steel pan and melt them together over moderate heat.

(2) When the beeswax and cocoa butter are just barely melted, add the shea butter to the mix, stir until melted, and reduce the heat. While cooling the mixture, spray the molds with mold release. When the mix cools to the point where you can barely pour it, stir in the peppermint oil.

(3) When well mixed, pour the liquid into the molds slowly and carefully, as close to the mold as reasonable to avoid splashing and bubbles in the cream later.

(4) Let the cream set in the molds for 24 hours at room temperature to cool and harden.

(5) When cool, remove the bars from the molds carefully so they aren't damaged. These types of creams are soft.

(6) Store in a sealable plastic bag to protect the volatiles and to keep the cream moist. Use sparingly—a little goes a long way.

CHAPTER 5 →25 Rules of Modern Beekeeping←

The first sections of this book cover the management skills you need to maintain healthy, productive honey bee colonies for your first couple of seasons, including basics of honey bee biology, colony management, and environmental factors. Now it's time to apply those skills with an eye towards reducing the impact of *Varroa* mites and keeping your bees as healthy and happy as possible.

Whether you have one colony or a hundred beeyards, some basic rules always prevail. One is that your bees need to eat and that good nutrition is essential. However, what they eat as larvae is not what they eat as adults, and what a pampered queen needs to eat is not what a hard-working forager needs to eat. But the rules extend beyond your bees' food. Even more necessary is a clean, safe, and steady water supply. And safe, appropriate, and adequate shelter is necessary at all times, too. It's easy to see a problem when the cover blows off or the supers are tipped over, but it is harder to determine if more room is needed for an expanding population and increasing food stores. Adequate protection from pests and diseases, though not quite as easy to manage, are at least easy

to diagnose and for the most part easy to treat. And not only should reproduction be on the list of considerations, but choices that improve each generation should be offered.

A honey bee colony is a complex thing to categorize. There are three types of individuals to consider, all of whom change how they react, physically and emotionally, to their environment as they age. Then there are the various subgroups. Additionally, like all politics, all beekeeping is local, so coming up with a list of all-encompassing rules is a challenge. My most common answer to almost every question regarding honey bees and beekeeping is, "It depends," so be aware of that as you apply these rules. There will certainly be exceptions.

These rules are divided into five categories and are not hard and fast. Use them as you see fit, as reminders or guidelines. If there's controversy about the order of importance, the first and second would argue the loudest. A colony cannot exist without a queen. But any colony will not exist infested with too many *Varroa* mites. Therein lies the difference: The queen's presence is black and white. *Varroa* is a gray scale that's always changing.

Queen's Rules

1. Queens must be raised in luxury.
2. Queens must be well mated before being sold or used.
3. Queens must be productive.

Queen's Rule Number 1: Queens Must Be Raised in Luxury

There's an old saying: We too often get the queen we can, rather than the queen we want. Getting the queen you want means finding a queen producer who provides queens of the quality you want, at the time and the price you want. Ideally, you will one day raise your own queens, suitable to your exact location, management style, and timetable. When those skills are acquired, this rule will no longer be needed. Still, a beekeeper should know enough about the process to ask good questions. So here are some of the basics of raising good queens, and what can go wrong in the process.

When just starting out, most of us get our bees in packages—2 to 4 pounds (1 to 1.8 kg) of bees and a queen. If this is the case, find out if the package producer's specialty is raising bees or raising queens. There is a difference. To get everything but the buzz out of their bees, some package sellers pollinate very early in the season and then produce honey later. Having strong colonies early in the season from a good pollination buildup means the colonies will swarm if the population isn't reduced. Shaking packages from these full-up colonies accomplishes that goal directly, and the producers get double duty out of the bees: They were pollinating a crop a week ago, trucked home, shaken into a package, and sold. Double duty, double money. True, many package producers also raise queens, but early queens can be iffy because of finicky weather, a dearth of drones, or a late return from pollination and a late start. Thus, some package producers simply raise bees, then add queens from a queen producer to the packages. So the queens in your packages may or may not be from the package producer, and they may or may not have been raised in the lap of luxury. It pays to ask.

Here there is one thing any beekeeper can do to increase the probability that the queen purchased was raised in luxury: Watch the weather wherever the new queen is being produced. It's pretty easy to find out, almost in real time, what the weather is anywhere on the planet from your smartphone or computer, and it is a tool you would be foolish not to use. Here's why.

A luxury childhood begins with a healthy breeder queen and colony. These are the bees that produce the larvae that include your colony's eventual queen. Inclement weather during very early spring can limit food. Cold can chill brood,

thus reducing the potential worker population in the colonies that raise your queen. A period of very warm weather earlier than normal can produce a flush of *Varroa* mites ahead of a treatment schedule, causing all manner of direct problems by injuring the bees that care for the breeder or the breeder herself, and it can definitely increase the virus load in the whole operation. Most queen producers work to keep all pest and disease levels low in their operations, but the virus load in a colony follows every individual. So, virus-infested attendants can pass it on to the breeder queen, to other nurse bees, to larvae, or to drones. Once it starts, there's no stopping it, only slowing it down.

If the breeder queen's colony is successful in keeping their queen healthy, she will produce eggs and larvae that will be moved to a different colony, one populated with an artificially large number of very young nurse bees with very active food glands. They are able to feed these soon-to-be queens with huge amounts of royal jelly, the food of queens. But they also decide whether a particular larva is fit to be a queen. They feed those deemed fit by their own set of rules, and they don't feed poor quality, injured, or unacceptable larvae. This colony, usually called a starter, can have weather or other problems, too. A cold snap can be devastating by chilling this expensive brood, poor quality food because of drought or freezing can harm the nurse bees, and too many *Varroa* and viruses can bring their own set of issues. These are the same problems other colonies could have, but control here is especially critical. After twenty-four or forty-eight hours in the starter, the now-accepted queen cell holding your queen is moved to yet another colony, called a finisher. These colonies continue feeding your queen and finish making the beeswax cell she will pupate in. Of course feeding is critical here: Poor stores or sick nurse bees can give your queen a bad start before she ever emerges to mate. So, when it comes to your queen being raised in luxury, the weather can play a significant role. It pays to pay attention.

But even when the weather is perfect, other problems can interfere with your queen's luxury requirements—sick or hungry nurse bees for instance. Sickness can nearly always be pinned on *Varroa*, past or present, and it is difficult to know if the queen producer is doing an adequate job of *Varroa* control. A related issue is chemical residue in the colonies your future queen is living in, whether it's the breeder queen colony, the starter, the finisher, or the mating nuc. All could have residue problems. The chemicals come from beekeepers, agriculture, homeowners, or natural plant toxins in the environment. And they soak into the wax your queen is living in. This is never good.

So the questions you should be asking are pretty simple: Are the queen's mother and all the nurse bees in her mother's colony healthy? How often are healthy, young nurse bees added to the starters and finishers? How much chemical residue is in the breeder's colony, and how often is old comb removed? If you've been watching the weather, you have an idea of food availability. What does the producer feed when the weather turns? Breeder queens should have the best of all worlds so she is producing the best of all offspring. Many queen producers finish the queen cells in a temperature- and humidity-controlled incubator, which, certainly, should be well controlled. Is it?

If all of these conditions are met—clean, safe, well fed, and under control—the queen you will eventually receive is off to a good start. But the game's not over yet.

Queen's Rule Number 2: Queens Must Be Well Mated Before Being Sold or Used

One or two days before she emerges, the queen's cell is moved from the finisher colony to a mating nuc, which is usually a much smaller colony, often only a two or three half-size frame nucleus colony. She takes a few days to get ready to fly, and the nurse bees take care of her. Then she has a short window of time to make those several flights. If the weather turns sour and flying isn't possible, or only possible for a day or two, your queen will not have the opportunity to meet and greet as many drones as is optimal. So even though she is mated, she is barely mated, and will run out of sperm much sooner than a well-mated queen. How many drones should she mate with? Twenty to twenty-five is good, maybe even thirty. But five or ten is a disaster. Watch the weather for the two weeks before your queen is to be delivered to make sure there is some sunshine and good flying weather. A poorly mated or unmated queen will be replaced by the colony, usually very soon after introduction, and you will lose valuable build-up time. It would be anthropomorphic to say that a well-mated queen seems more confident than a poorly mated queen, but it's a good way to imagine what's going on.

Poor mating can also result when the drones your queen mates with perform less than spectacularly. Two problems can cause this: damage by *Varroa* or the chemicals used to control *Varroa* and damage due to poor nutrition. The most important question to ask is, how many drone colonies does the queen producer support to produce enough drones for all those queens to mate with? Drone colonies have drone comb added especially to produce extra drones. If each queen needs twenty-five drones, and the producer is selling thousands of queens a week, drone demand will be heavy. Do they have enough?

To be fair, *Varroa* is endemic and queen producers are between a rock and a hard place because of this beast. They have to protect their drone-producing colonies using chemicals, exposing their drones to those chemicals, which can be damaging. Chemical exposure means a shorter life span, flight issues, sperm production issues, and general health issues. Some producers don't treat drone colonies for that reason, but then they should have more drone colonies. It's a tough balancing act.

Nutrition, of course, is a constant issue as well. All manner of things can limit food supplies—late or early springs, increasing development, farming practices. And if natural sources aren't available, the producer needs to supply the food directly. Early in the year, most colonies are supplied with artificial carbohydrates such as corn syrup and protein supplements anyway. But weather and timing can get in the way. If the weather was bad, what did the producer do?

So, if your queen was able to take mating flights at will and was able to find many healthy drones in the drone congregation areas and mate with fifteen to twenty, maybe more, from a variety of genetic backgrounds, she is well mated.

She should remain in the mating nuc for a period of time so the producer can evaluate her. This could be only long enough to see that she is laying eggs, long enough to see that she is laying a solid pattern, or long enough to see that her offspring actually emerge. But the longer she stays there, the more the producer has invested, and the more she will cost. What is the price of two expensive but poorly mated queens, compared with one expensive, excellent queen? Ask your producer.

Queen's Rule Number 3: Queens Must Be Productive

This rule is probably the most subjective of all of the twenty-five rules. What is productive, anyway? Well, here it means that she makes you enough money, honey, bees, pollination contracts, pollen or propolis, or wax. Your management style, business goals, and location all shape the way you measure how your bees perform. Your best metric is comparing your queen's productivity with another queen's productivity. If another colony is producing much more honey than yours, why? Location, number of bees in the box, health, interference from the beekeeper? When you know what you want, you have to find the queen that produces the bees that get you there. It's that simple, and that complicated.

One metric to use will answer some of these questions. How many eggs is she laying every day? At the height of the season, it is estimated that she should be laying about 1,800 to 2,000 eggs a day. It's not impossible to calculate. Here's the quick way: After she has been introduced and has been laying for at least two weeks and you are satisfied with her pattern and see a healthy retinue of nurse bees around her, count the number of cells of sealed brood. That's not as hard as it might sound. You can measure the area of sealed brood on every frame using a ruler to get the area in square inches. Though it varies some, count on 25 brood cells per square inch (6.45 sq. cm). That will give you the total number of eggs she laid during the twelve days since all the cells were capped—some yesterday, some twelve days ago. Divide that total by 12 and you have an average eggs-per-day figure. Do it again in twelve days to see if it changes up or down. Soon you will have a good idea of area at a glance. Figure a deep frame has about 4,500 cells on a side and a medium has about 2,700 cells per side. But don't guess. Count the number of cells vertically and horizontally, multiply, and then you'll know. Estimate the percentage covered in sealed brood and calculate from there. Does she measure up? Is she getting better, or slowing down?

Your main goal, then, is to identify what you want. When you know, you'll also know if your queens are productive. If they aren't, replace them.

Bees' Rules

1. Your bees should be adapted to your location.
2. Your bees should be selected to your management style and technique.
3. Your bees should be resistant to pests and diseases.
4. Your bees should be well behaved.

Bees' Rule Number 1: Your Bees Should Be Adapted to Your Location

Most of the queens in the United States are raised in the Southeast, South, or far West from early spring to midsummer so that you can get them when the traditional books say you should have them—about the time of fruit and dandelion bloom where you live—and in time to requeen summer splits and colonies gone queenless from swarming. This used to work. There is a better way.

Obtain or raise a queen from stock that's been surviving and thriving where you live, not in the balmy, humid South or warmer or drier-than-usual West. A third or fourth generation stock that produces a virgin queen where you live has managed to figure out bloom timing, the weather, and the peculiarities of your region. Of course you need her to mate with drones with the same experience. If this is possible, you will produce queens that produce bees that thrive where

they live. Resistance to *Varroa* and other pests may or may not be part of the package, but adaptation is high on the list. Once you have a line that likes where they live, you can begin selecting for other traits you want. But first, they have to stay alive.

Bees' Rule Number 2: Your Bees Should Be Selected to Your Management Style and Technique

If you pollinate for a living, your bees need to wake up early to take advantage of the first nectar flows of the year (or your early-season feeding) to be ready for those early crops. If they sleep in and wait until spring has settled and life is easy (no matter how much you feed or stimulate them), you are going to have a problem. However, if you are always late in getting started in the spring, they would be just the thing.

If honey production is your goal, you want a colony that has a very high population of bees just before the main honey flow so they can take advantage of the large number of foragers available to gather nectar. If you want a line of bees that can survive Arctic winters, they must have small populations that eat much less food that a larger colony and do not begin raising brood until there is ample food coming in.

Choose wisely when deciding on a stock to ensure that it matches your style of keeping bees.

Bees' Rule Number 3: Your Bees Should Be Resistant to Pests and Diseases

Developing a breeding program when you are just starting out or have only a very few colonies is difficult and probably impossible. So your second choice is to seek locally produced queens who produce stock that suggests some resistance to the common problems or to find more distant sources that have the traits you want. However you obtain such stock, resistance or tolerance to *Varroa* and other pests and diseases is definitely important to have. Make this a priority.

There are several avenues of resistance to pests and diseases that breeders work with; however, most of them use some form of the honey bee's natural hygienic behavior. This can include simple grooming: removing adult *Varroa* mites from each other or vigorous removal of infested larva or pupae from cells, either sealed or still open. There are other mechanisms you can select for, but these are the most common to find, measure, and incorporate into a breeding program. The choices become complex when looking for the best stock to raise—local, resistant, productive, and gentle—and choosing the right one depends mostly on what your operation needs and doesn't have, and what level and kind of IPM or chemical treatments your program calls for, all the way from none to very disciplined.

There's an argument for not selecting for hygienic behavior because it can become detrimental to the colony if the behavior is too aggressive. Rather, some breeders are selecting for different pest and disease resistance traits and at the same time selecting for longevity of the queen. A queen that meets and maintains your metric of productivity is necessary, of course, but if her line demonstrates very low swarming behavior (so she stays around), and she maintains her productivity level for three or even five years, you will be far ahead of the game. Long lived and resistant—to date those queens are rare, but can be found if you look for them.

Bees' Rule Number 4: Your Bees Should Be Well Behaved

There are obvious reasons to keep gentle bees: If they sting you a lot, they are not easy to work with, and if they are not easy to work with, you won't take care of them. And, aggressive bees in your backyard are a danger to your family and your neighbors.

That should be the end of the story. But Nature happens. Sometimes a colony is a kitten one week and a tiger three weeks later. Several things can produce this. Population increase simply means there are more bees around. A dearth could initiate very protective behavior from robbing. Skunks and other creatures could be harassing your colonies, keeping them defensive.

Selection for foraging behavior may preclude guarding behavior. Often a very high tolerance of alarm pheromone resides in the population, such that no matter how much a colony is disturbed and how much a few guards react, the majority of the bees fail to respond and you have a "gentle" colony. A gentle colony is easier to work, is less troublesome in a crowded setting, is less prone to rob other colonies, will not follow you, or will only follow you for a few feet (about 1 m), and will not fly in your face. The bees tend to not run on the comb, fewer or none fly when a colony is opened, and they react dramatically submissive when smoked. All this makes it easier and faster to get into a colony, do what needs doing, and be done. This is the way it should be.

Breeders use selection criteria for "gentle" that range from being able to open a colony without using smoke to wave a hand over the top bars to work colonies during inclement weather to kick colonies, and more. Queen lines deemed well behaved, that is gentle, continue. Extreme selections have been made over the years toward producing gentle bees, and records of beekeepers working bees in swimsuits on a cool, cloudy day without being subjected to stings or aggressive behavior of any kind are common.

But all bets are off when the carefully selected queen in this colony is lost and the colony raises a new queen that mates with any number of unknown drones. A too-common belief is that mean bees make more honey. True or false? Often, it's true because the colony superseded the gentle queen and the replacement mated with a few of the local boys. Some are from gentle colonies, but some will be from feral colonies that have survived because they are definitely not gentle. Plus a local colony knows the local environment—weather, forage, timing, wintering—better than a colony that is spending its first summer in the same location. These bees almost always produce more honey than those imports from balmy climates. But are they gentle? Probably not.

So a colony that was a kitten a couple of weeks ago may turn on you the next time you visit. And these new traits tend toward survival rather than beekeeper-friendly behavior. And productivity is definitely a survival trait. Perhaps now you can see even more the value of marking a queen.

Beekeeping Rules

1. Be aware of the rest of the pests.
2. Keep your wax clean.
3. Isolate your bees from other bees.
4. Avoid agriculture, all the time, at all costs.
5. Provide enough room for bees and brood.
6. Provide enough room for nectar and honey.
7. Manage swarming.
8. Make sure there's enough good food all the time.
9. Remove weak and diseased colonies and combine small but healthy colonies.
10. Keep excellent records, take good care of your equipment, and have extras.
11. Take care of the bees that take care of the bees that go into winter.
12. Winter appropriately.

This group isn't listed in any priority order because they all come in first. They are all critical to successfully raising your honey bees and keeping them alive. Nor are they listed in any calendar schedule; for most of these rules, you need to pay attention all season long.

Beekeeping Rule Number 1: Be Aware of the Rest of the Pests

Considering the many things that ail honey bees, you would think this section would be quite long. But if you know about these problems, you can easily find prevention, treatment, and recovery information, all of which will differ depending on where you are. That said, without doubt *Varroa destructor* (and the virus complex it harbors and spreads) is the King of Bad in a beehive, but the other pests and diseases cannot be ignored or the fate of the colony will always be the same—dead, dying, or declining—certainly NOT thriving. With the exception of American foulbrood, a colony can handle most pest and disease problems if it has a healthy, large population with a queen who produces offspring that exhibit some level of hygienic behavior but are not overly zealous about it, have enough good food all the time, live in clean wax, are situated in full sun, and are able to avoid constant, though possibly occasional, contact with agricultural pesticides. This is as basic as it gets for IPM. Next up are mechanical tricks: traps for small hive beetles, proper super storage for wax moth, good ventilation, and ongoing selection for bees that thrive when confronted with these problems. There are chemical means of controlling all of these pests, even American foulbrood, and you may choose to go that route when time and colony numbers dictate.

Get rid of old wax. Melt it down and replace with new, clean foundation.

But American foulbrood is different in that there's no escaping the longevity of the spores it leaves behind. Once infected, the bees' home is condemned. You can apply an antibiotic to save the bees, destroy with fire the frames and inner cover, if you have one or the cover if you don't, and scorch the insides of the boxes to destroy any spores. A safer choice is to destroy the bees also, but some choose to go to the effort of saving the bees and putting them on new, clean equipment. It's time consuming, but most often it's worth the effort considering the cost of replacement.

Whatever choices you make, you must be able to recognize these problems, implement remedial action—traps, for instance—and know what to do next if preventative actions fail. Be aware of all of the problems, expect them to arrive, and know how to save your bees.

Beekeeping Rule Number 2: Keep Your Wax Clean

This has always been a rule, but it has become especially important now that beekeepers are using chemicals for *Varroa* control. Wax absorbs these chemicals and, as a result, your bees are constantly exposed to low levels of them, leading to stress at best and pest immunity eventually. But even if you aren't using in-hive miticides, your wax still becomes laden with toxins from the outside world your bees routinely visit. Plants produce natural toxins to protect themselves from their pests and some toxins return to the hive with bees. Too, there

are the agricultural toxins your bees encounter on occasion. Add to that the daily grunge that comes home when bees are gathering water in ponds and puddles, walking on soil and the propolis, cocoon, and frass buildup in each cell, and eventually, it's time to clean house. When comb is dark brown to black, or generally has been in the hive about three years, it should be replaced. If you use hard *Varroa* chemical controls, replacing every other year might be better.

Beekeeping Rule Number 3: Isolate Your Bees from Other Bees

This is much easier said than done, but the fact is that you are your neighbors' beekeeper, and if they are less fastidious than you when caring for their bees, your bees will inherit or steal their problems. Drifting especially enables problems to spread, but abandoned equipment rife with disease also is a problem. And there's a fine line between offering to help a fellow beekeeper do a good job and being a pain in the neck and interfering. Moreover, one beekeeper's perfectly acceptable, and legal, management style is a fussy beekeeper's continued source of irritation. It's not easy being clean, or green, and isolation is almost impossible. But try anyway.

Beekeeping Rule Number 5: Provide Enough Room for Bees and Brood

The normal cycle of population fluctuation in a honey bee colony is generally smallest in late fall, gradually but slowly increasing as winter progresses, building faster as spring approaches, fastest and most in late spring, plateauing in mid- to late summer, decreasing slowly during autumn, and then back to almost no growth in late fall. Of course numbers are arbitrary depending on where the bees are, how much forage is available during any one season, and how they are or are not managed.

A honey bee colony managed for optimum production without stressing the colony or demanding more than should be asked needs space to accommodate the growth of both juvenile and adult bees, pollen and honey in the broodnest area, and of course the hoarded stores that will get them through lean times ahead.

Providing all that space all the time can be problematic as there will be more space than the small population of the colony can safely protect from predators—wax moths, ants, roaches, small hive beetles, and the like—so additional space must be added just before it is needed. This is timing at its best.

This involves knowing how much space today's eggs will require when they become adults in three weeks and what to continue to expect from a queen who can lay upwards of 1,500 to 2,000 eggs a day. But you have to know that she really does lay that many—how can you tell, and what does it mean (see Queen's Rule No. 3) All those bees have to have some place to be, and if there's not enough room, they'll take matters into their own hands, and then you have to worry about swarming (see Beekeeping Rule No. 7).

You can make a quick calculation of the space that you need by looking at the amount of sealed brood (you're doing that anyway because you are counting eggs per day from the queen). Look closely at a frame of sealed brood. A worker bee standing on that frame covers two cells. So, when you figure out sealed brood numbers, you double that space to accommodate the adults that will emerge. Suddenly you need way more room than you thought, and if the queen's in the increasing phase, that number will double again in about three weeks. Plan ahead.

Beekeeping Rule Number 4: Avoid Agriculture, All the Time, at All Costs

If you have any choice at all, keep your bees as far away from industrial agriculture as you can—miles and miles away if possible. Modern food production has developed technologies that poison the plants, flowers, nectar, pollen, soil, and groundwater everywhere we produce food. Not enough to outright kill you, your bees, or the rest of all of the wildlife that mostly doesn't exist anymore in those regions, but to permanently affect them with sublethal and barely lethal doses. Your bees carry these toxic groceries back to the hive and feed them to their young. This weakens the hive, lowers individual bees' immune systems, adds another layer of stress on an already overtaxed system, and as a final kicker, gets in your beeswax and continues to dole out its deadly dose for years to come. Stay away from farms.

Additionally, as industrial agriculture increases in size, diversity in the landscape—and a bee's diet—decreases dramatically. So your bees must travel farther to maintain a balanced diet, or they simply don't get enough good food all the time. Herbicides, no more fence rows, and infrequent alternate cropping systems have reduced plant diversity to a bare minimum.

Beekeeping Rule Number 6: Provide Enough Room for Nectar and Honey

Along with the bees needing room in a hive, expansion plans must always be ready to accommodate incoming nectar and the resulting honey. As a rule of thumb, nectar is about 70 percent water and 30 percent sugars and solids, while honey is about 83 percent sugars and solids and 17 percent water. The range of sugar in nectar goes from about 10 percent to as much as 80 percent, but 30 is a good average to consider and what bees seek as the lowest concentration to chase.

How much room? If your colony is to make 125 pounds (56 kg) of honey from 30 percent sugar nectar, they will need to gather right about 300 pounds (136 kg) of nectar, or about 75 gallons (284 L) of nectar, to get 125 pounds (56 kg) of honey. That's a significant amount of required room, and even though not all at the same time, if it's not there when needed . . . production stops.

To get from mostly water to mostly sugar takes a bit of magic, chemical transformation, and room. Returning foragers hand off their collected nectar, already laced with the enzymes needed to change the disaccharide sugars (mostly sucrose) to the monosaccharide sugars (mostly fructose and glucose, but a host of others in small amounts) to house bees ready to finish the process. They, too, enhance the mix with enzymes, manipulate the droplet so it is exposed to the warm internal environment of the hive, and then hang a drop or two from the top of an empty cell, giving it maximum exposure to the drying winds in the hive and dehydrating it even further.

When that drop is dry enough, it is moved to a cell that will eventually be full of honey and then covered with a thin layer of wax to protect it. Some of this honey is also placed in cells about half full of stored pollen to protect the pollen. These cells are not covered in wax.

The critical task for a beekeeper is making sure there is adequate room for all those dehydrating drops. If house bees determine there's no more room for incoming nectar, their enthusiasm for taking those nectar loads from incoming foragers drops off the edge of the comb. In a way, the message is, don't bring any more sister, there isn't room, at least for now. So, rather than go out and get more, those foragers take a break. And every break is that much less honey collected.

Those drops don't remain long in those cells—overnight or maybe a day—before they are reduced to storage quality and moved to a honey cell, opening up that cell for more incoming nectar. But that time, whether it's just a few hours or a whole day, is lost to honey production. So if you figure there's a super full of locust nectar out there this year because the bloom is spectacular, there's no sign or rain or cool temperatures, and you've got lots of bees, make certain they have at least a super for nectar storage, and two would be good insurance. You can remove the extra space when the locust flow is over.

Beekeeping Rule Number 7: Manage Swarming

Swarming is a complex behavior—see more about it on page 148. The indications generally given are that the hive is crowded and the weather is good. It's more than that, of course, but that's a start. In the spring, the colony is expanding at a rapid rate as the population increases by more than a thousand bees a day, up to two thousand if everything is going well (allowing for natural mortality, these numbers are from 750 to 1,500 bees per day). So there simply needs to be room for all the adult bees in the colony. But when it is clear there isn't going to be enough space—a great deal of brood, lots of adults and a rich cache of incoming food—the trigger is pulled and the process begins. The queen's diet is curtailed so she slows then stops laying eggs and loses weight so she can fly, and food storage, too, slows to make room for all the new bees. There is a measurable break in the brood cycle when a colony swarms, which does have positive effect on *Varroa* life cycle. A critical component here, however, is the concept of queen pheromone per bee.

As queens begin to age past the second or third year, the colony-uniting pheromones they produce begin to change. There are fifteen, maybe twenty different compounds they emit that hold the colony together, and as a queen's capability diminishes, she begins to have less of an effect on the bees' behavior. Too, the ratios of these compounds change, allowing some behaviors to erupt, while still keeping others in check. Crowding, lots of available food in the real world outside, time of year (increasing day length rather than decreasing day length), population, and an aging queen all contribute to the process. Unchecked or uncorrected, a healthy colony will swarm, just as it is supposed to do.

Managing that behavior is time consuming, labor and equipment intensive, and expensive. Not managing it is simply expensive in lost bees and reduced crop. So the easiest though not necessarily most efficient, productive, or bee-friendly approach is to reduce the population by making a split or two, or even three, from the original colony long before it reaches the point of making a swarming decision. Stop it before it starts. Then, provide a new queen for both the mother colony (led by an older, swarm-proven queen) and the split (so you know what you will have rather than what you hope you will have). Divide and conquer is one way of looking at it, and replace the throne in all of the now new colonies made from the original.

Dividing a large colony into several smaller colonies can result in less honey, lost queens, absconding, and slowness to build. But ending the season and going into winter with several new, honey producing, healthy colonies led by young, vigorous queens can also be the result. Whatever your choice, you have to manage swarms, or you will lose colonies.

Beekeeping Rule Number 8: Make Sure There's Enough Good Food All the Time

This sounds simple: Feed your bees when they need food. It isn't that simple, unfortunately. Actually, it's feed your bees BEFORE they need food. You need to know when to begin building up your colony's population to reach a peak right before a honey flow starts so they are not building on the flow but maximizing storage. You should know when honey flows are about over, when the next one starts, how much food is stored, and how much brood needs feeding for how long. The nurse bees should have access to as much food as they want when feeding young, or they begin to use the protein from their own bodies, which shortens their lives, and if prolonged, shortchanges the young they are feeding. Anticipate food shortages before they happen so when they should happen, they don't. Use the formula that it takes a cell of pollen plus a cell of honey plus a cell of water to make a bee—in summer that's 1,000 cells-plus for each bee, each day.

Beekeeping Rule Number 9: Remove Weak and Diseased Colonies and Combine Small but Healthy Colonies

Don't waste time with runts. A small, unproductive colony costs as much maintenance time as a large colony, and, on a per bee basis, costs ten to twenty times as much to maintain. Determine why colonies that aren't sick aren't keeping up. Most likely it's the queen's performance. Unless it is early, early spring, dispatch that queen and combine that small colony with a stronger healthy one, or at least one of similar strength, to get a bigger, more productive colony with more bees able for forage. You have reduced your workload, increased your harvest, freed up equipment for other bees, and simplified overwintering.

Determine why sick colonies are sick. Diseases such as the foulbroods and nosema, parasites such as small hive beetle, *Varroa* and its unseen viruses, and tracheal mites can all be dealt with. Sometimes, and often, dealing with them means dispatching the bees, and even the equipment if infested. As with healthy, small colonies, the time and energy spent per bee on sick colonies is enormous, the cost of drugs is high, and the fact that you are perpetuating a line of bees that succumbs to any of these diseases should be seriously considered.

In short, don't waste time with runts.

Beekeeping Rule Number 10: Keep Excellent Records, Take Good Care of Your Equipment, and Have Extras

Good records will save you money, make you honey, keep you from losing things, help you find things, keep your spouse and family happy, save bees, and keep you from pulling your hair out. So will taking good care of your equipment.

Good record keeping can be accomplished in any number of ways: a spiral or three-ring notebook, a bound notebook, a recorder in the field that you speak into, a smartphone or notepad, electronic devices kept permanently in the hive—the list goes on and on. That's the easy part. The harder part is getting your records into a form they can be used. Keeping them in the cab of your truck is not very useful. Writing on or using bricks on the tops of hives or, even worse, on the inner cover, is not very useful. Leaving the book in the garage is not very useful. You see the point.

And here is the best piece of advice I can give you regarding record keeping. It's simple, easy and extraordinarily efficient: Give every colony a number. Paint it on the cover with spray paint and identify it in your records by that number. Have a page in your book or a file in your computer for each colony. When that colony expires, and it will, start a new colony in it, remove the data sheets in your book or file regarding the previous colony with that number, and begin again. You can always find the sheet with the number twenty-one. You can never find the sheet that says, "third from the left, when facing south." You can expand this to beeyards if you manage by the yard instead of by the colony. The Barnes yard, for instance, on a date, received syrup. This gives some good information, but loses some of the detail. If every colony in a yard is a split that receives a queen cell, and you get only 20 percent take, what went wrong. Or, if you get 95 percent take, what went right? And what about those that didn't take? Put in a frame of open brood and hope? Colony notes on beeyard scale are less informative, but on that scale, you manage by the yard, not the bee.

Many beekeepers have field notes and house notes. They take notes in the field—on paper, on a recorder, in a book. Speech-to-text recorders and phones are easy to use and are getting more accurate all the time. Then they put those notes, and even expand on them, on a computer or in another notebook when the day is over. This accomplishes several things: You are forced to relive the day and in so doing, you will recall things that didn't get written down. This will enable you to make up a To-Do list you can take with you next time—supers needed, feeding needed, check queens, requeen, weed control, check *Varroa* sticky boards, medication needed, replace broken covers, bring material to make new hive stand so that apiary can be expanded, remove drone comb *Varroa* traps, and other tasks. Without the To-Do list, how many of these get done? Or, how many get done after you make a second trip back to the shop?

Locations, queen source, weather, time of day, temperament of the bees, sounds, sights, unusual, usual—everything can get recorded. Right along with this is having a hive scale. Recorded at the same time every day, you can estimate a hundred things going on in your hives—honey flows especially, so you know what's happening with the rest of your bees.

Take care of your tools and they'll take care of you; it's as true here as in a machine shop or auto repair garage. Keep boxes, tops and bottoms, protected from the elements. They're only wood, and they live outside most of the year. Keep your smoker cleaned out, and clean your hive tools every time you use them (a bucket of water, dish soap, and a metal scrubbing pad will last you most of the season and work just fine). Keep your suit, gloves, and bee veil tight and clean (a dirty suit with lots of venom invites a bad day).

Boxes and frames used strictly for honey production can be cleaned pretty well when uncapping and extracting. It adds time, but keeps things fitting together. Burr, brace comb, and especially propolis build up over time and things just don't fit right. Power wash after extracting, and separate them into piles by good riddance, repair, repaint, and okay. If the timing isn't good, then consider a cleaning session later in the year when boxes aren't needed and can be examined, cleaned, and made ready. One beekeeper I know always has a small sledge hammer in his truck. When a piece of equipment has worn out, he smashes it, right there in the field. That way, it is absolutely no longer useful and won't get saved to still be broken in the future.

Have extras. I always tell beginning beekeepers to buy two, better yet three, hive tools right off. You'll leave one in the beeyard or drop one in the grass or leave one at home when going to the beeyard. Same with brushes, smokers, and the like. As you gain experience, and beeyards, you'll find that leaving smoker fuel, a hive tool, a brush, and maybe extra frames in an old super with a cover is a good idea. It'll save a lot of travel. But more importantly, if you absolutely need something—your truck for instance—what do you do when you don't have it? Having two is best but often impractical. Having a friend with one is second best. If you can count on that, you have a chance to save the day. Have extras of everything, and have a Plan B if you don't.

Beekeeping Rule Number 11: Take Care of the Bees That Take Care of the Bees That Go into Winter

Think of it this way: If your grandparents are working at less than full speed, they can't take good care of your parents, who are raised in less than ideal conditions, and in turn can't take good care of you. Bees damaged by *Varroa*, the viruses, nosema, or malnutrition at any stage of the game are not hitting on all eight cylinders, and they can't care for those depending on them as well as they should. Nor can they forage, clean, feed, or care for the queen; nor do they live as long as they should. So the hurt continues each generation, becoming worse and worse.

The final chapter is that bees dying young leave a hole needing to be filled, so just-born bees begin foraging at younger and younger ages, trying in vain to support the queen and feed the young. But the damage done renders them damaged, and they fly away and do not return. You find an empty hive with brood, food, and a queen . . . but no bees. Or, it happens later in the winter, and come early spring, it's the same thing: No bees in the boxes, and you wonder what happened.

What happened is that *Varroa* got the best of them way, way back in the middle of summer. It started then: The viruses got hold and there was no turning back. Treatments for *Varroa* in the fall killed *Varroa*, but there were only dead bees walking in that hive by then. The grandparents were damaged, the parents were damaged, and their children are damaged. The hive is doomed.

Organic acid Varrao *treatments are safe for bees, wax, and give good control.*

Beekeeping Rule Number 12: Winter Appropriately

Where your bees are makes a difference. If you live in the tropics, you will need to make preparations for a continuous nectar flow, with room for storage, young bees to be raised, and the like. In more moderate climes, where there is a cold or perhaps dry time, but not a devastating winter, there may be many days bees can fly and forage, and brood is constantly present. Here, too, room must be available, but at the same time, if there's a prolonged cool period, ample food must be available. And for some climates, those that are mild but barren, food is the key ingredient. The weather may be perfect, but if it's a nonblooming desert, your bees will starve even faster because they will fly and find nothing. So some winters aren't cold—just hungry. It's the cold winters, though, that are the hardest. Long stretches of nonflying weather, with the bees huddled together inside, means that food, warmth, and room are always problems.

Honey bees are tropical insects, and although evolution has equipped them with a hoarding instinct to preserve food for lean times and the ability to form wintering clusters, winter is still a struggle.

And even these traits don't ensure success. Several factors must be in place to get from late fall to late spring. First, there has to be enough food. How much? The range is somewhere between 50 pounds (22.7 kg) of honey in the warmer temperate areas where some foraging is possible, to 150 pounds (68 kg) where winter is seven to nine months long and subfreezing temperatures are routine.

And there always needs to be a minimum of five frames of honey in a hive to keep everybody fed—spring, summer, fall, and winter. And there needs to be enough pollen—enough to feed an increasing population as spring approaches. Research indicates there should be about 500 square inches (0.3 sq. m) of stored pollen for a strong, healthy hive to do well over the winter. It can also be in the form of the internally stored protein in overwintering bees. There needs to be a healthy amount of this, commonly called fat bodies, but there must be more so that when feeding the emerging population of young bees, these workers don't completely decimate their internal stores and threaten their own longevity and survival.

Next, there needs to be enough bees to occupy the space in the hive and some empty cells that bees can occupy. There also needs to be empty space to hold enough bees between

frames, too, and enough bees to get from one part of the hive to another without becoming separated from the rest. A group of bees needs to be able to go around the edge or top or bottom of a frame to get to the other side or another frame for the food stored there. Once bees reach that food, they can pass it back to the group in the middle without becoming stranded.

Some beekeepers—and in the past many beekeepers—make temporary passages in frames by punching a hole in the center for the bees to easily pass from one frame to another and from one side of the box to the other without losing touch with the cluster of bees. Some even remove a center frame to allow a large cluster to form, and open a passage in several frames so bees can easily pass from the center to the edge to find stored food without having to cross the Rubicon of frame edges. This arrangement in wild hives and even top bar hives is common, made in the fall and repaired in the spring to make life easier for the bees.

But beekeepers tend not to want to damage comb, and if plastic comb is used, opening spaces isn't feasible, although removing a comb would be. And maybe it should be. The biggest obstacle is that most hives, if not most beekeepers, move to the south each winter to take advantage of honey flows there, to make splits and get ready for early spring pollination contracts.

To compensate, in cold regions beekeepers provide protective measures to mediate the internal temperatures, and especially the internal temperature changes during the course of a day. These include external wind breaks made of bales of straw, evergreen boughs, temporary fences, snow fences—anything that deflects winter wind around or over the hives to keep the colds blasts from directly hitting them.

But where it's really cold, additional protection should be used, and this, unfortunately, has fallen out of favor for a couple of reasons. The first is that there is pressure to select

those bees that winter the best in extreme cold, with little food and small populations—certainly admirable traits. Why should we, as an industry, prop up a population of bees that is not resistant to *Varroa* or other diseases, suffers and dies in harsh environments, and exhibits other not survival-in-my-climate attributes? This is an ongoing debate, and there are strong arguments for both sides. The live-and-let-die approach is, in the long run, productive, but the carnage-strewn beeyards attest to the cost in bees, money, time, and energy.

The other reason is technology and time. It is easy and inexpensive to simply pick up your bees and move them to a warmer place. This has become standard for most commercial-sized operations, and is growing even with smaller operations. One ambitious person with access to transportation can move 400 to 450 colonies in one trip and settle them in in a warm location. The bees do better, the beekeeper has some relief, and that ambitious truck driver has a job. Another technique is indoor wintering—controlled to varying degrees for ventilation, temperature, CO_2 levels, and mites, but for a shorter period of time, like the potato storage caves used in the American West.

But until this winter-over debate is resolved, winter protection is one good way to give bees the edge during harsh winter weather. These measures range from a simple roofing paper wrap to a lightly insulated plastic coated wrap to a simple weatherproof heavy duty cardboard sleeve that slips over the colony and folds over the top, encasing the colony completely, or at least the top half or so. The next step is a heavy duty insulation blanket that, like the box, slips over the colony. Additional protection can be added by joining colonies together in groups of four on a pallet and treating them all as a unit, wrapping them together so that each has two inside walls protected from the elements by another colony. The protection this offers is amazing.

All of these techniques have one issue in common: ventilation. When warm, moist metabolic air rises through the colony, it eventually reaches the top of the cavity, which is usually the inside bottom of the inner or migratory cover. What happens then can save or kill a wintering colony. If the inside top of the hive is cold because there is no top insulation, that warm air will condense when it hits the inside surface and form liquid water. That water simply drips down on the bees, chilling and killing them. Or, it can collect on the top bars of the top super, the comb, even the top bars of the

super below, and freeze into a road block of ice, not allowing the bees access to the honey stored there. Come spring you will find nearly a super full of honey in the top, the dead bees below having starved because they couldn't reach the honey covered in ice. And, of course by now the ice is gone and you are left wondering why the bees simply didn't move up when they ran low on food down below. Dumb bees, or dumb beekeeper?

However, if adequate ventilation is provided, that warm, moist air rises and escapes from inside and causes no harm at all. Insulate the top of the hive even better than the sides, and that warm air never cools and escapes, causing no problems at all. Moreover, that ventilation duct provides a top entrance for the colony when snow and ice accumulate and block the lower entrance.

In locations with less extreme weather, good ventilation is always a good idea, but protection beyond a wind break is often not necessary.

In some places, wintering colonies in a building—or in the case of the western United States, a former potato storage shed—works well as long as internal temperatures are sufficiently cold to keep bees clustered, but warm enough to allow movement to food and not extreme stress. This has proven to be very effective for short term storage between the onset of winter and the movement for almond pollination. Longer term indoor storage requires careful monitoring of temperature, CO_2 levels, circulation for good oxygen supply, and so on. The list goes on, but the results are staggeringly successful for those with enough colonies to make this efficient. Winter can, indeed, be a pleasant place to be if it's in the right place.

The bottom line for wintering appropriately is to provide more than enough protection, continue to select for lines of bees that do well in your winter location, follow the basic principles of physics relative to moisture and condensation, and have enough good food for all winter long.

Varroa Rules

1. Learn about *Varroa*.
2. The best way to control *Varroa* is the worst way to make honey.
3. Use IPM treatments first, soft next, and hard never.

Varroa Rule Number 1: Learn About Varroa

It is well known that the *Varroa* mite, *Varroa destructor*, switched gears a few decades ago and moved from its primary host, *Apis cerena*, often called the Eastern Bee native to southeastern Asia, to our domesticated European honey bee, *Apis mellifera*, which is in use in much of the rest of the world.

Apis cerena live in cavities like our bees, are a bit smaller and generally much more aggressive, and are more like their distant African cousins in that they produce lots and lots of bees that swarm and swarm and swarm rather than put that energy into honey production. Some selections have been made by beekeepers over the years, and there are lines of *A. cerena* now that are less aggressive, swarm less, and produce enough honey to make keeping them productive. For the most part, however, *A. cerena* are unmanageable, aggressive, and extremely competitive when forced to share an environment with *A. mellifera*.

A quick review of *Varroa*'s life cycle is instructive here. A pregnant female *Varroa* mite enters a honey bee colony on the back of a visiting bee, on a bee moved there by a beekeeper, or on brood frames shared between infested and uninfested colonies. This pregnant female searches for a larva that's ready to spin its cocoon and have the cell capped by the nurse bees to protect it while it changes from larva to adult (honey bees have a typical complete metamorphosis cycle). The female mite heads to the bottom of the cell and hides beneath the larva so it is not detected, and when the cell is capped it is inside the cell alone with the larva. She prefers drones over workers because drones take longer to evolve from larva to adult, and she can lay more eggs and produce more young while protected in the cell, but if there are no drone cells available, she takes a worker cell in a heartbeat. As soon as the cell is capped, she begins feeding on the larva for a protein meal, lays a male egg, continues feeding, and lays another egg, then another if time allows. During this very short time, the male matures and mates with his sisters. When the larva leaves the cell, with it leave the original female and one or more pregnant females, ready to continue the process.

The initial host–parasite relationship that developed over eons has struck a balance between *A. cerena* and *Varroa*: Simply, the parasite agreed not to completely kill an *A. cerena* colony, and the *A. cerena* agreed to let the mites kill off some of the drones as long as they left the workers alone. And, since swarming is a big part of the *A. cerena* cycle, the *Varroa* never had a good opportunity to build up to large numbers in an individual colony before—poof!—everybody was gone, and it's time to start all over again. Because of these two behaviors, that agreement still works.

Now, enter well-intentioned people from parts of the world where *A. cerena* doesn't exist. They bring bees that don't swarm nearly as much, are far less aggressive, make much more honey, and are bigger and take longer to develop—*A. mellifera*.

As soon as *A. mellifera* was introduced to *A. cerena* regions, *Varroa* saw an evolutionary advantage to feeding on this new host. And they did with devastating results. Because of the longer development time, the pregnant female mite had time to lay more eggs before the adult bee emerged from the cell and mite reproduction increased significantly in an *A. mellifera* colony. Moreover, because *A. mellifera* didn't swarm nearly as often, the mites had more time to build up huge populations, thus constantly damaging both larval and adult bees—and now both drones and workers—and after sometimes one, usually two, maybe as many as three years, there were so many mites the colony would abscond or collapse.

This is not a profitable host–parasite relationship; both mites and bees lose because they simply have not had those eons of time to work out an agreement so both survive.

It didn't take long for *Varroa* to gain a foothold in every *A. mellifera* colony brought to the area, and once established on *A. mellifera*, and with no negative reports from its association with *A. cerena*, infected *A. mellifera* bees were moved all over the world with *Varroa* mites in tow. Now the only place these mites are not found is Australia.

Part of the *A. cerena/Varroa* agreement must have contained something about transmitting viruses, since that doesn't seem to be an issue with these two. But with *A. mellifera*, *Varroa* infestations compromise our honey bee's immune system to almost everything, but especially diseases and viruses. Moreover, when a mite feeds on a honey bee infected with one or more viruses, those viruses are transmitted first to the mite, and from the mite to another bee. And it gets worse: That infected bee can pass along those viruses to juvenile bees when feeding, from queens to eggs, from workers to drones, and from worker to worker. *Varroa* mites have unleashed the perfect storm of colony destruction, and after decades and decades still remain the worst offender of beekeeping continuity there is.

At first, mites always won and *A. mellifera* colonies always died. Then, chemicals came into play that would kill mites and not bees (killing a bug on a bug), but even though the chemicals didn't kill bees, they weren't kind to them either, and the beeswax these chemicals came in contact with simply soaked up this poison, and still, mostly the mites would win. But after almost 100 years of trying to work it out, some *A. mellifera* colonies began to handle the problem.

Evolution and compromise are both slow in the insect world. But experience does pay, and those longest exposed, *A. mellifera* bees from eastern Russia, gained an upper hand and began to show resistance to *Varroa* mites, and the two, as with *Varroa* and *A. cereana*, began to be able to live together.

This is where we are today with *Varroa*. Some *A. mellifera* bees have some resistance. Most don't, and there is no major organized program to breed resistance into the population. Rather, any breeding done in that direction is done by smaller-size, more focused and dedicated beekeepers and queen breeders. Profit is in bees that die, not live, and profit is in research about a problem that kills bees. If bees lived, nobody would be able to make a living selling them, or researching *Varroa* mites.

This, then, is *Varroa*. Know it well, fear it, and strive to control it. And there are ways to do this without chemicals.

Varroa Rule Number 2: The Best Way to Control Varroa Is the Worst Way to Make Honey

The more brood there is in a colony, the more opportunity there is for *Varroa* to reproduce and build its population. And, since a pregnant mite can produce about 2.5 *Varroa* mites in drone cells and about 1.2 mites in worker cells—that's more mites than bees—lots of brood means lots and lots of mites. Just when the honey bee population in the hive reaches a profitable population relative to a summer honey flow, so, too, does the *Varroa* population, exactly at the time that drone populations begin to decline. So what's left for the next generation of *Varroa* mites? Worker brood. The scenario here is pretty clear: *Varroa* outnumber workers, the *Varroa*/virus complex escalates, and soon most bees in a hive have been infected with a virus—usually several. This reduces life spans and keeps nurse bees from being fully able to care for the young, foragers from being able to compete at 100 percent, and the remaining drones from being competitive in mating. The colony is in trouble and, in all likelihood, collapses because younger and younger bees become foragers to feed the steady supply of brood, but the foragers are already sick, die young, or fly away and don't return.

There are some lines of bees available that are mostly tolerant or resistant to *Varroa*, the viruses, or both, so they might tolerate a larger mite population than most other colonies. These lines, however, very often depend on several factors to maintain that status. Mite pressure is one; this is

one of the main reasons to keep your bees as far as possible from other colonies. Low or no mite pressure certainly makes tolerance easier. Few or no other pests or diseases, a variety of nutritional sources on a continuous basis, water, winter protection, swarm management, a healthy and productive queen, minimal beekeeper interference, and a location in full sun keeps a colony strong, viable, and able to exert all of its resistant or tolerant properties. Shortchange one or more of these and the walls slowly begin to crumble and *Varroa* gets a foothold.

But while it is desirable, even envious, to have bees that handle mite populations, no matter how they do it, many don't, and the mite population builds to unmanageable levels. Beekeeper input is required to reduce that population. And here's where the real world of mite control and honey production come head to head. To avoid mite populations building, you stop them before they start, or you interfere with the buildup after it gets going.

A strong overwintered colony with few mites (you are checking mite populations, right?) is your best bet because it has the fewest bees and lowest amount of brood. If the colony is strong, a split and a treatment then will knock the population right down to almost zero. Leave the splits queenless (or confine the queen from the original and don't introduce, or release, a queen into the split) for a brood cycle (three weeks), and all that remains are exposed mites with no place to go. Monitoring will support this, but even a moderate population of bees can remove many—sometimes all—of these unprotected mites. After the broodless period, the queens are released and the colony can expand. But the colony has missed a three-week growth spurt—three weeks without any new bees introduced right at the time the colony needs new bees the most. That's bad for honey production. But, no mites.

Another technique is to introduce treatments at this time of year. Placing drone comb traps and using soft chemicals will reduce mite populations enough for the colony to build, but the mite population will also continue to build. Monitoring the mites then becomes more important. When levels reach a detrimental point, which will be on the upside of the colony's growth curve, a late spring or early summer split will help. Of course this will provide a setback for your honey production. But each split now has half the mites the parent colony did, and requeening, or at least stopping brood production for a cycle, will dramatically reduce mites while the colony makes some honey. Treating each right now adds another layer of protection, certainly. A soft treatment with honey supers on

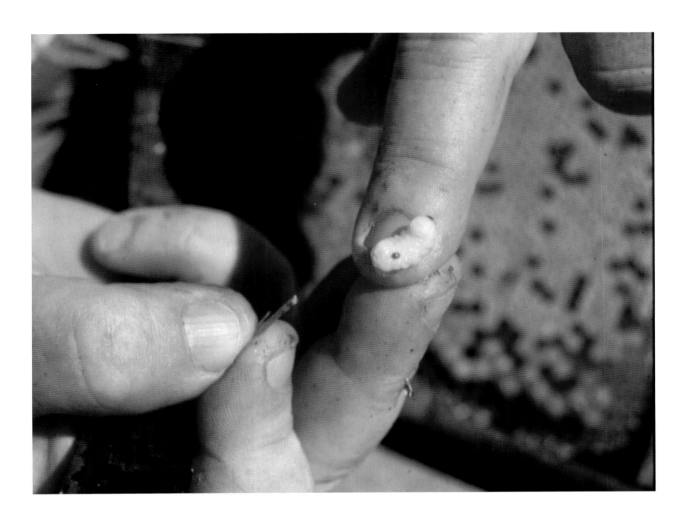

will stop many of any remaining mites, and no brood slows those remaining even more. Requeen after a brood cycle, replace honey supers if removed, and your colony is ready for a fall flow and a healthy winter. This technique is effective if your area has a slow period after the early spring flows and before the heavier summer bloom begins—sometimes called a summer dearth or a June gap. It's a window of opportunity for the colony's cycle to be intentionally interrupted but reduce the honey crop the least. It is also labor and time intensive.

Varroa Rule Number 3: Use IPM Treatments First, Soft Next, and Hard Never

Realistically, beekeepers with many colonies seldom have the time or help to carefully monitor individual colonies and manage each independently. There'd be just too much record keeping. Many make management decisions on a beeyard scale, which is better than managing by the calendar, but the calendar still has a vote because honey flows and swarms get a vote, too. And to stay in business, *Varroa* and its demon virus complex have to be controlled. On a commercial scale, a beeyard is a manageable unit. Some percentage of colonies from each location are tested, and the whole yard is treated for the level of infestation detected. This means each yard, whether ten, fifty, or more colonies, is treated the same. The next yard may not be treated at all. The next treated twice. It all depends on the results of the test. So test.

It's the testing here that's important. Treatments are expensive in labor, material, and stress on the bees, and avoiding a treatment is the first, best choice. Time of year

is also important, and if there's a honey flow on, many treatments are not allowed or the honey has to be pulled so a treatment can be applied.

From an IPM perspective, isolation, full sun, drone trapping, sugar dusting, eliminating nutritional stress, making sure you have strong colonies with healthy and productive queens, and eliminating other pests and diseases are the best choices. Second, then, are the few chemical treatments that can be used with honey supers on. Next are the organic acid treatments because they leave no residue in the wax and are relatively easy on the bees; next, the essential oil compounds. And that's it. Even the oils leave residues. The harder chemicals leave toxic residues, and those are to be avoided. Period.

Beekeeper Rules

1. Seek continuing education.
2. Know all about beekeeper safety.
3. Food safety isn't the last rule, it's the first.

Beekeeper Rule Number 1: Seek Continuing Education

I have yet to meet a beekeeper who knows it all. Some are pretty good, but nobody, absolutely nobody, gets it right all of the time. And just when you think you do, something changes. Something new, something different. Something.

These are fundamental points, but they are so often neglected that they bear repeating.

When starting out, read, read, read. Not just the new beginner's books. Get the classics that are still around. There's lots that's new in the last 100 years, but there's lots more that's the same.

Take a beginner's class. Find the best one you can. Take two if you can. Every instructor has strengths and weaknesses, and you'll gain from both.

Find a mentor, someone willing to help you learn. Work for free for the experience. You will never learn as much sitting in a class as you will watching, then doing, then doing again with the guidance of a skilled and experienced beekeeper.

Then, take the beginner's class the second year. You won't believe how much more you will learn.

Join a local beekeeping association so you know what's happening nearby today.

Join a regional association so you know what is going to happen nearby sometime soon.

Join a national association so you know what the government has planned, what the newest pest is and the treatment for it, and what is going on more than five states away that will eventually get to your state. Attending the meetings is often difficult, but web pages, newsletters, and webinars will keep you up to date on the issues that will put you out of business, kill your bees, or make you a criminal if you don't stop doing some of the questionable things you've been doing.

Learn what they call advanced skills, even if you never have more than a couple of colonies. Learn the biology and techniques of the many ways to rear queens. Take an artificial insemination class. Try them all. Make your own equipment. Make customized equipment. Learn to make and sell nucs with your local queens.

Volunteer to work with a beekeeper who does things you don't do. Move bees for pollination. Shake packages. Make comb honey or liquid honey. Try top bar hives.

Try new things as often as possible. New equipment, new techniques, new people.

When you travel to someplace new, look up a beekeeper or an association of beekeepers. Go to their meeting. Work their bees. Take your spouse with you to help or just to meet another beekeeper. Finding local associations is easy on the Web, and from there a beekeeper is right around the corner.

Teach a class. You'll never learn more than when you have to teach. Queen rearing. Making splits. Making equipment. How to use a solar wax melter. Have a field day at your home and show how you do things to people who do them differently. Do part of the annual beginner's class. Hold a special advanced class and bring in an advanced beekeeper to help.

Take one of the many online courses available. Even the beginner's classes that you thought you already knew.

Buy at least one new book a year. Read it. Get the new DVD and see if they do it right.

Try this at least once: Make up a super colony. Fortify it with lots of bees and brood and food, and see how much honey you can make with a super colony. Go for a record.

Learn how to make honey with a two-queen colony.

Move bees at night, with someone's help.

Never quit trying something new.

Beekeeper Rule Number 2: Know All About Beekeeper Safety

There are a lot of ways you can get hurt when dealing with bees, some obvious, some not so. Here are a few of the major ones.

Stings

Of course bees sting, and stings around the face can cause permanent injury. When working bees, wear a veil. Bee suits help prevent stings but serve more to keep you clean. However, sometimes bees have to be worked when the weather is bad, at night, or after they have been harassed by harvesting, moving, being split, skunks, a bear, or other problems. Defensive, protective behavior can border on downright aggressiveness then. Wear a veil, suit, and gloves. Seal the cuffs. Cover rips and holes with duct tape. Enduring the pain of a sting or two is one thing, but lots of stings can render you ill, or worse.

Lifting

Ergonomics is too often overlooked, especially when just starting out. If you spend your days lifting, bending, carrying, pushing, and working hard, you already know how to lift, how not to twist, and to get help when needed. But most of us tend toward computer and couch time, and working out is more often limited to an occasional game of golf, tennis, or a jog in the gym.

Beekeeping isn't like that. It's bending and stooping and lifting boxes that weigh in at 40 to 100 pounds (18 to 45.4 kg). And you are lifting them, if you set them on the ground instead of on a stand, anywhere from 2 to 4 feet (0.6 to 1.2 m). And often you turn to one side, pick up the box, lift with your back and not your legs, then rather than take a sideways step, twist to set the box back on top of the hive. So you've thrown out your back and pulled every muscle from your hips to your shoulders. Have fun at that computer on Monday.

Robbing

Honey bees are opportunists. If they find a source of food close by rather than far away, they'll take the closer food. If they find one with lots of sugar compared to not much, they'll take the lots. If they have more kids to feed than they have food, they'll find it somewhere. Put some combination of these together and you can initiate a robbing situation in your apiary. Strong colonies will rob weaker colonies. Early compared to late foraging times can also lead to robbing. All manner of things can start a robbing event.

Once begun, robbing escalates rapidly, and just as rapidly, safety deteriorates. The colony being robbed is protective, biting and stinging the robber bees trying to gain entry. The robbers sting back. Most of this happens on the landing board outside the colony. Soon the entire apiary is filled with alarm pheromone. A slight breeze wafts it into your front yard, then on through the neighborhood. The rest of the colonies in your apiary are now on alert, and bees from those hives a block away are on alert. A dog walks by and gets stung, then the man walking the dog, then the kids next door. You can see where this is going in a rapid, and perhaps deadly, manner.

So avoid setting up a robbing situation. Keep colonies approximately the same strength. Small or weak colonies should have the smallest possible opening, which is easier to defend. Don't work colonies during a dearth, which sends the smell of honey to foragers from all colonies. Don't leave colonies open while you go back to the garage to get the tools you forgot. And don't throw bits of wax and comb on the ground when working a colony during these times. Give bees in the apiary no clue there is food nearby when no food is available anywhere else. They will find it, and they will start robbing.

Harvest time can be particularly troublesome. There are some guidelines to follow no matter which method you use to harvest that can increase your safety and that of your family and neighbors. If possible, the day before you harvest, go to the beeyard and loosen every box you are going to harvest from. Simply lift it up so one end and both sides are separated. This breaks all burr and brace comb between boxes, along with all the propolis bonds. The bees will clean it overnight, and the next day the box or frames will be much easier to move, and more importantly, there won't be liquid honey exposed to incite robbing.

If you are removing individual frames because the super is too heavy to lift, there are some precautions you can take to reduce the tendency to rob, which can be high in the fall when there are fewer nectar sources available and lots of bees are home. Have replacement frames ready (drawn comb best, foundation next) to replace those removed. Then quickly remove a capped frame from the super while still on the hive, gently brush bees off at the front door and not back into the super so you are removing bees rather than concentrating them in the super. Be gentle so you don't break a lot of capping and place the frame in a bee-tight box. When done, replace the honey frames with the empties you brought with you.

If removing entire supers, once the bees have moved down, remove the fume board from the top of the super. Then lift the super off the hive, and place it on a flat board or an overturned cover you have prepared ahead of time. Immediately cover it with another cover or flat board to keep out any curious bees. If you are using the fume board for another hive, place it on that hive right away so it can start moving bees down while you finish. Or, place it on the super below as soon as you have the first one off. If using an escape board, remove the cover and inner cover, then the super, leaving the escape board on the super below for the moment. Place the super on your prepared, overturned cover or flat board and cover immediately with another flat board. You can then put the escape board on another hive, or below the next super, and return tomorrow to repeat the process.

When finished extracting, let bees clean out the wet supers (boxes that have been extracted but are still sticky with honey residue) either far away from the apiary or directly on the colony the super was removed from.

But what do you do when your prevention actions fail and a robbing situation starts, it gets out of control, and you are in the middle of what can truly be a dangerous, perhaps deadly, event?

Control Strategies

Control strategies range from being very aggressive to simply making a few changes. The first and most important thing to do is protect the colony(s) being robbed. Replace the covers, block the entrances, stop any and all opportunity for conflict between robbers and robbed. You can put screens in the entrances if you have them, but that's not often the case. Rather, use entrance reducers, grass, rags—anything to completely seal the entrances. If there are several colonies robbing, consider turning the tables: Remove the covers and inner covers of every colony that is robbing. This suddenly puts them on the defensive, and they need to protect rather than attack. In extreme cases, placing a lawn sprinkler on the apiary may help. The confrontation between robbers and robbed has to be stopped or it will spread. Leave the robbed colonies blocked overnight if possible, and when opening, provide as small an entrance as possible.

The Ultimate Challenge

The worst case scenario is that a colony or colonies become completely out of control, stinging, harassing, and endangering everything within reach. At some point the decision may be made that this colony is a danger and must be destroyed. This is certainly more common in urban settings than country beeyards, but be prepared anyway.

Here's how. When you know that a colony must be destroyed immediately, do not debate the decision, and do not delay. Act now!

Have on hand two medium-size dishwashing detergent containers, two five-gallon (19 L) pails, and enough water to fill them.

Fill each pail, and mix one container of dish soap into each. Wearing protective gear, completely block the entrance of the colony. If using a screened bottom board, insert the winter block so the bottom is solid. Quickly remove the cover and inner cover and pour the contents of the pail into the hive and replace the cover only. Wait five minutes and pour the second pail in. In five minutes or less, almost every bee in that colony will be dead or dying. You have solved a serious, dangerous problem.

A second way is to have a large, plastic garbage bag always handy. Slip the bag over the offensive colony, lie it down, and tie the bag shut at the bottom—problem solved, even quicker than with soapy water.

Health Issues

These include some things you might take for granted that you shouldn't, such as good lifting practices. Know the symptoms of heat exhaustion and stroke, and always have water with you, even if you are twenty yards from the kitchen. Hot days, no breeze, and a full bee suit can lead to overheating and serious illness. Know, too, the first signs of an allergic reaction. It happens, even to beekeepers with years of experience. Watch out for shortness of breath, itching, hives, or dizziness. Have a sting kit handy. You'll need a prescription—they are not inexpensive. Funerals are more expensive. And take your cell phone with you. Every time.

Does someone know where you are when you go to an outyard? Do you have a map? Did you tell someone so they know where to look when you don't come home for supper?

Beekeeper Rule Number 3: Food Safety Isn't the Last Rule, It's the First Rule.

The honey, pollen, propolis, and even the wax you harvest from your colony are food products. At all times you must treat them as such. Use as little smoke as possible when working in honey supers. Make certain you do not use chemicals in your hive when honey supers are on so the honey super wax doesn't come in contact with it. Exchange brood comb every two to three years to keep all residues— beekeeper applied, farmer applied, and nature applied—to a minimum.

When harvesting with fume boards, use as little as possible for as short a time as possible to avoid tainting the honey. When moving honey supers, keep them completely covered so dust and debris cannot enter between the frames.

In the honey house, don't warm the honey higher than 100°F (38°C) before uncapping, if you warm it at all. Keep your uncapping area clean, your uncapped frame area especially clean, and your extractor cleaned between uses. Do not let small hive beetles damage honey in supers, so extract

Nothing beats hot water and soap to keep the honey house clean.

as soon as possible. Store extracted honey in clean pails, and make sure your final container is clean before filling. If using a bottling tank, do not leave any honey in it between bottling sessions. Clean any pipe filters between bottling sessions. Let extracted honey sit for at least a day before bottling it so any small particles can rise to the top and be removed.

Clean pollen before selling using a blowing machine or, if time allows, picking out nonpollen pieces. Store in a tightly sealed container or freezer before selling.

Allow bees to clean supers of residue honey before storing, and store unused supers to allow light and air between frames, while using excluders or screens to keep mice out.

✦Conclusion✦

Like almost every endeavor, the harder you work at bee-keeping, the better you'll become. However, working hard is not necessarily the same as working smart, and too often the two don't come together. In fact, by not working smart you almost always have to work even harder, and you still may not reach your goal.

Working smart requires knowing what works, what doesn't, and why. Experience will help—including learning from mistakes you or other beekeepers made in the past. The full value of working with an experienced beekeeper even before you begin cannot be measured. That time spent learning will pay dividends long into the future.

The premise of this book is threefold: First, to explain in detail what a good honey bee colony should be (knowing what works). Second, to examine the things that will probably go wrong at some point (what doesn't work), and third, to show that things come along you didn't expect, but are now prepared to successfully deal with based on your understanding of how honey bees work (the why).

For nearly three decades I have produced a monthly magazine covering every aspect of honey bees and beekeeping. It has given me a unique perspective on what information beginning beekeepers need most, first. Every year the same questions come across my desk. These include spring beginnings with package bees or nucs, summer management that covers honey production and colony health issues, then harvesting and honey processing questions, then on to preparing for winter, and finally early and late spring management for overwintered colonies, which absolutely must cover swarm management and biology. That's the seasonal cycle of a honey bee colony and a road map for a beekeeper's responsibilities.

I answered all the questions I'd received in the first edition of *The Backyard Beekeeper*. Then CCD arrived, and the second edition came to be. Now here we are in the third edition, adding to the mix the newer issues of honey bee nutrition and agricultural pesticides. This edition still answers the questions that have needed answers since the first beekeeper put bees in a container. But keeping bees has changed in the last five years, so it also answers those questions never before asked. And now our pages cover some aspects of keeping bees that even anticipate the future and what to expect.

But of course answering a summer management question on the phone is much more difficult and much less fulfilling than being there and showing and doing. So my goal has been to show how to do things as much as possible. Not using glamour shots—I've tried to make the photos straightforward and easy to understand. Lots of photos make it easier to comprehend and at least approaches being there in person.

But neither honey bees, nor the environment they live in, are static. These forces are constantly changing with the weather, the external interference of urban development, undiscovered pests, and the constant barrage of danger from agriculture in general. The information and instructions in the front of this book have been updated to recognize this changed world. In fact, much has changed since the previous edition of this book. This is why this time we have provided "25 Rules For Modern Beekeeping." These rules are so fundamental that I believe they will not change over time, though the means to meet them might change.

Though the rules themselves won't change, the importance of any one of them will wax and wane over time. And new rules may come to be, due to the changing world our bees will live in. If you are prepared, changing rules won't bother you.

Staying ahead of the needs of a colony is much easier than playing catch up all season; or replacing colonies that perish because of neglect; or worse, not knowing what to do or when to do it.

Another significant change between the first and third editions is *Varroa* management. We spend a lot of space in this edition dealing with this monster. *Varroa* have not gone away, but the information presented this time will help you handle the problems they create. I encourage you to spend time on that section. Putting poison in a beehive is one of the dumbest things we can do, and it is to be avoided at every opportunity. Hygienic behavior, drone trapping, and especially using bees resistant to this demon mite are all possible. But if treatments of some kind become necessary, the tools we have now are a bit more benign than they once were. Essential oils and volatile organic acids are effective and leave the least poison behind to contaminate the wax, harm the bees, and dissolve into the honey.

Since the last edition, the problems with queens have risen to be among the top three issues needing our attention. The best way to get a good queen is to raise her yourself, but that's often not in the cards. However, if you intend to keep on keeping bees, raising your own queens is the ONLY way to get quality, local, survivor stock. You probably won't do this during your first year, but next year take a class and try, and the following year take the class again and just do it. Read a few books, work with a queen producer or breeder, and start selecting the type of bees you need in your operation.

The term *Colony Collapse Disorder* has been around for several years now, and we've learned much about this slippery beast. It describes a combination of several issues that are relatively new in the bee world. That they came together isn't surprising because they are all related. These include *Varroa* mites and the viruses they spread, *Nosema cerenae* and the damage it can cause, poor nutrition due to expanding monocultures in agriculture creating less and less dietary diversity, and constant exposure to agricultural chemicals including herbicides, insecticides, fungicides, in-hive mite control agents, and antibiotics applied to the hives by beekeepers. All of these are bad by themselves, but together the synergy they build exceeds anything beekeepers have experienced. And in general, the situation has not improved much. Bees still don't have enough food, the weather has been erratic for a decade, poisons keep coming down the road, and resistant bees are almost impossible to find. The end of this story still remains to be told.

It can appear overwhelming when viewed in its entirety. But when you break it into smaller pieces, each becomes a manageable chore. And now we can do just that. We know about enough good food all the time. We can provide bees locations that are free of agricultural influences, we can manage *Varroa* without toxic chemicals, and we can obtain honey bees that resist all of these dangers.

Keeping honey bees remains one of the most interesting, productive, and challenging hobbies or businesses that exists. I've been at it for more than thirty years, and I still learn something new every season, and I'm still having fun in the process. You will too. Enjoy the bees.

A modern queen-breeding operation

Glossary

A.I. Root—founder of the first and once the largest beekeeping equipment manufacturing company in the U.S., located in Medina, Ohio.

Abdomen—the third region of a body of a bee enclosing the honey stomach, intestine, reproductive and other organs, wax and Nasonov glands, and the sting.

Abscond—the action of all bees leaving the hive due to extreme stress, disease, pests, or danger, such as a fire.

African honey bees—a subrace of honey bees, originally from Africa, brought to Brazil, that has migrated north to the United States. They are extremely defensive and nearly impossible to work.

Alarm pheromone—pheromone released by worker bees during an emergency.

American foulbrood (AFB)—a brood disease of honey bees caused by the spore-forming bacterium *Paenibacillus* (formerly *Bacillus*) larvae.

Anther—the part of a flower that produces pollen; the male reproductive cells.

Apiary—where honey bee colonies are located; often called beeyard.

Apiculture—the science and art of keeping honey bees.

Apiguard—a thymol-based gel applied to honey bee colonies to reduce infestations of *Varroa* mites.

ApiLife Var—a thymol-based liquid soaked into florist foam and applied to honey bee colonies to reduce infestations of *Varroa* mites.

Apis mellifera—the genus and species of the honey bee found in the United States.

Apistan—a plastic strip impregnated with fluvalinate (a toxic pesticide) that is inserted into honey bee colonies used to kill *Varroa* mites.

Bait hive—a box, often an old brood box, composed of a comb or two, a top and bottom, and a small entrance hole, used to attract swarms. It is often placed in an apiary.

Balling the Queen—The action of worker bees attacking a new queen, or a queen cage, intent on killing her because she is foreign. Often occurs during queen introduction.

Bee bread—a mixture of pollen and honey used as food by the bees.

Bee escape—a device used to remove bees from honey supers during harvest by permitting bees to pass one way but preventing their return.

Bee space—¼" (0.6 cm) to ⅜" (1 cm) space that bees live in.

Beeswax—a complex mixture or organic compounds secreted by eight glands on the ventral side of the worker bee's abdomen; used for molding six-sided cells into comb. Its melting point is from 144° F (62° C) to 147° F (64° C).

Bee veil—a cloth or wire netting for protecting a beekeeper's face, head, and neck from stings.

Bee venom—the poison secreted by glands attached to a bee's stinger.

Beeyard—a location where honey bee colonies are kept.

Bottom board—the screened floor of a beehive.

Brace/burr comb—comb built between parallel combs, adjacent wood, or two wooden parts such as top bars.

Brood—the term used for all immature stages of bees: eggs, larvae, and pupae.

Brood chamber—the part of the hive in which the brood is reared.

Capped brood—pupae whose cells have been sealed as a cover during their nonfeeding pupal period.

Cappings—the thin, pure wax covering of cells filled with honey; the coverings after they are sliced from the surface of a honeyfilled comb when extracting the best beeswax.

Carniolan—dark-colored race of bees from Eastern Europe, which are very gentle.

Caucasian—grayish-colored race of bees from Europe, use excessive propolis.

Cell—a single hexagonal (six-sided) compartment of a honey comb.

Chalkbrood—a fungal disease of honey bee larvae.

CheckMite—a plastic strip impregnated with coumophos (a very toxic pesticide) inserted into honey bee colonies used to kill *Varroa* mites.

Chilled brood—developing bee brood that have died from exposure to cold.

Cleansing flight—a quick, short flight bees take after confinement to void feces.

Cluster—a group of bees hanging together for warmth.

Colony—adult bees and developing brood living together including the hive they are living in.

Colony Collapse Disorder—a pathogen-driven condition in a honey bee colony in which the adult bees are stricken and leave the hive to die. Ultimately only the queen, a very few young bees, and (depending on the time of year), a large quantity of brood are left. There is an abnormal amount of time before secondary scavengers move in—wax moth, small hive beetles, or robbing bees—and if bees are put back on the abandoned combs immediately, they, too succumb to the problem. It is suspected to be virus-driven, but concrete proof of that has not been discovered as yet.

Comb—a sheet of six-sided cells made of beeswax by honey bees in which brood is reared and honey and pollen are stored.

Comb foundation—a commercially made sheet of plastic or beeswax with the cell bases of worker or drone cells embossed on both sides.

Comb honey—honey produced and sold in the comb, made in plastic frames and sold in round, plastic packages.

Compound eyes—a bee's sight organs, which are composed of many smaller units called *ommatidia*.

Cremed (Crystallized) honey—honey that has been allowed to crystallize under controlled conditions.

Cut-Comb Honey—Comb honey, cut-comb honey, and chunk honey are all derived from a full frame of capped honey, made on foundation without wires. Cut-comb honey is a piece of comb honey cut to fit into a container. The edges are drained before being packaged. Chunk honey is a piece of cut-comb honey placed in a jar, which is filled with liquid honey.

Dancing—a series of repeated movements of bees on comb used to communicate the location of food sources and potential home sites.

Dearth—a time when nectar or pollen or both are not available.

Dividing—partitioning a colony to form two or more units, often called divides or splits.

Drawn comb—comb with cells built out by bees from a foundation.

Drifting—bees going to a colony that is not their own.

Drone—the male honey bee.

Drone comb—comb measuring about four cells per inch (2.5 cm) in which the queen lays unfertilized eggs that become drones.

Drone layer—a queen able to produce only unfertilized eggs, thus drones.

Egg—the first stage of a honey bee's metamorphosis.

Encaustic Painting—a painting technique in which color pigments are added to melted beeswax, which is used like oil paint for painting or drawing.

Entrance reducer—a wooden or metal device used to reduce the large entrance of a hive to keep robbing bees out and to make the entrance easier to defend, and to reduce exposure to wind and the elements outside.

European foulbrood (EFB)—an infectious brood disease of honey bees caused by the bacterium *Melissococcus* (formally *Streptococcus*) *pluton*.

Extracted honey—liquid honey removed from the comb with an extractor.

Fanning or scenting—worker bees producing Nasanov pheromone and sending it out to bees away from the colony as a homing beacon.

Feeder—any one of a number of devices used to feed honey bees sugar syrup including pail feeders, inhive frame feeders, hivetop feeders, and entrance feeders.

Fertile queen—a queen that can lay fertilized eggs.

Forager—worker bees that work (forage) outside the hive, collecting nectar, pollen, water, and propolis.

Formic Acid—a chemical treatment for *Varroa* and tracheal mites. Pads of absorbent material are soaked in a strong solution of formic acid which are added to a beehive. The acid volatizes and the fumes are toxic to mite populations.

Frame—four pieces of wood/plastic (top bar, a bottom bar, and two end bars) designed to hold foundation/drawn comb.

Fume board—a rectangular frame, the dimensions of a super, covered with an absorbent material such as cloth or cardboard, on which a chemical repellent (Bee Go or Bee-Quick) is placed to drive the bees out of supers for honey removal.

Fumidil-B—one trade name for fumagillin; a chemotherapy used in the prevention and suppression of nosema disease.

Granulation—the formation of sugar (glucose) crystals in honey.

Grease patty—a mixture of vegetable shortening and granulated sugar placed near the brood area for tracheal mite control.

Guard Bees—after bees have been house bees, but before they become foragers, many spend time as guard bees: stationed at the front door or other entrance, checking incoming bees to make sure they belong to their hive.

Gums—sections of whole tree trunks with a complete, natural honey bee hive inside. These are then moved to a beeyard and are tended like a man-made beehive. The word comes from the tree species often preferred by bees: gum trees in the Appalachian region of the United States.

Hive—a man-made home for bees.

Hive tool—a metal tool used to open hives, pry frames apart, and remove wax and propolis.

Honey—a sweet material produced by bees from the nectar of flowers, composed of glucose and fructose sugars dissolved in about 18 percent water; contains small amounts of sucrose, mineral matter, vitamins, proteins, and enzymes.

Honey flow—a time when nectar is available and bees make and store honey.

Honey stomach—a portion of the digestive system in the abdomen of the adult honey bee used for carrying nectar, honey, or water.

Hygienic Bees—honey bees that have a genetic trait that pushes them to remove dead, diseased, or mite-infested larvae from a beehive are said to be hygienic.

Hymenoptera—the order of insects that all bees belong to, as do ants, wasps, and sawflies.

Inner cover—a lightweight cover used under a standard telescoping cover on a beehive.

Italian bees—most widely used race of honey bees in the United States; originally from Italy.

Langstroth hive—our modern-day, man-made, moveable frame hive named for the original designer.

Larva (plural, larvae)—the second (feeding) stage of bee metamorphosis; a white, legless, grublike insect.

Laying worker—a worker that lays drone eggs, usually in colonies that are hopelessly queenless.

Marked queen—Some queen producers sell queens that they mark with a spot of paint on the top surface of the thorax (the middle of three chief divisions of an insect's body). This makes the queen much easier to find, and indicates whether the queen you have found is the one you introduced or a new queen. Always use marked queens.

Mating flight—the flight made by a virgin queen when she mates in the air with several drones.

Metamorphosis—the four stages (egg, larva, pupa, adult) through which a bee passes during its life.

Nectar—a sweet liquid secreted by the nectaries of plants to attract insects.

Nosema—A digestive disease of honey bees, treated with Fumigillin.

Nuc or Nucleus (plural, nuclei)—a small, two- to five-frame hive used primarily for starting new colonies.

Nurse bees—young bees, three to ten days old, that feed and take care of developing brood.

Ocellus—simple eyes (3) on top of a honey bee's head. Used primarily as light sensors.

Package bees—screened shipping cage containing three pounds of bees, usually a queen, and food.

PDB (paradichlorobenzene)—crystals used as a last resort as a fumigant to protect stored drawn combs against wax moth.

Pesticides—the general name for chemicals used to kill pests of many varieties. Subcategories of pesticides are insecticides (which kill bees) and fungicides (which kill fungi, but can also be detrimental to honey bees). Combinations of insecticides and fungicides can be extremely deadly to foraging honey bees.

Pheromone—a chemical secreted by one bee that stimulates behavior in another bee. The best known bee pheromone is queen substance secreted by queens that regulate many behaviors in the hive.

Pollen—the male reproductive cells produced by flowers, used by honey bees as their source of protein.

Pollen basket—a flattened area on the outer surface of a worker bee's hind legs with curved spines used for carrying pollen or propolis to the hive.

Pollen trap—a mechanical device used to remove pollen loads from the pollen baskets of returning bees.

Pollination—the transfer of pollen from the anthers to the stigma of flowers.

Proboscis—the mouth parts of the bee that form the sucking tube and tongue.

Progressive provisioning—Not common in the insect world where adults feed their young continuously from egg eclosion to completed development up to pupation rather than mass provisioning, where all the food a larva will need for development is provided at one time.

Propolis—sap or resinous materials collected from the buds and wounds of plants by bees, then mixed with enzymes and used to strengthen wax comb, seal cracks and reduce entrances, and smooth rough spots in the hive.

Pupa—the third stage in the metamorphosis of the honey bee, during which the larva goes from grub to adult.

Queen—a fully developed female bee capable of reproduction and pheromone production. Larger than worker bees.

Queen cage—a small cage used for shipping and/or introduction of a queen into a colony.

Queen cell—a special elongated cell, in which the queen is reared. Usually an inch (2.5 cm) or more long, has an inside diameter of about ⅓ inch (0.8 cm), and hangs down from the comb in a vertical position, either between frames or from the bottom of a frame.

Queen cell cup—A round, cup-shaped structure that workers build on the bottoms of frames to accommodate a future queen cell. The current queen must place an egg in the cup before the workers begin building the rest of the queen cell. Queen cell cups are built most often just before swarm behaviors begin.

Queen excluder—metal or plastic grid that permits the passage of workers but restricts the movement of drones and queens to a specific part of the hive.

Queenright—a colony with healthy queen.

Rabbet—a narrow ledge on the inside upper end of a hive body or super from which the frames are suspended.

Requeen—to replace existing queen with new queen.

Robbing—bees stealing honey, especially during a dearth, and generally from weaker colonies.

Royal jelly—a highly nutritious glandular secretion of young bees, used to feed the queen and young brood.

Russian Bees—a line of honey bees that had spent generations exposed to *Varroa* mites without miticides. They were brought to the United States from eastern Russia for their innate resistance to mites.

Scout bees—foraging bees, primarily searching for pollen, nectar, propolis, water, or a new home.

Skeps—a woven basketlike container, often covered with mud or dung, used to house honey bees.

Small hive beetle—a destructive beetle that is a beehive/honey house pest living generally in the warmer areas of the United States. Originally from South Africa.

Smoker—a device used to produce smoke, used when working a colony.

Solar wax melter—a glass-covered insulated box used to melt wax from combs and cappings.

Spermatheca—an internal organ of the queen that stores the sperm of the drone.

Sting—the modified ovipositor of a honey bee used by workers in defense of the hive and by the queen to kill rival queens.

Sucrose—principal sugar found in nectar.

Super—a hive body used for storing surplus honey placed above the brood chamber.

Supersedure—a natural or emergency replacement of an established queen by a daughter in the same hive.

Surplus honey—honey stored by bees in the hive that can be used by the beekeeper and is not needed by the bees.

Swarm—about half the workers, a few drones, and usually the queen that leave the parent colony to establish a new colony.

Swarm cell—developing queen cell usually found on the bottom of the frames reared by bees before swarming.

Terramycin—an antibiotic used to treat European foulbrood. Also used for American foulbrood prevention, but it is not effective in killing the spore stage of this disease.

Thorax—the middle section of a honey bee, that has the wings and legs and most of the muscles.

Tracheal mite—*Acarapis woodi*, the tracheal-infesting honey bee parasite.

Tylosin—one of several antibiotics used to treat, but not cure, American foulbrood.

Uncapping knife—a specially shaped knife used to remove the cappings from sealed honey.

Uniting—combining two or more colonies to form a larger colony.

***Varroa* mite**—*Varroa destructor*, a parasitic mite of adult and pupal stages of honey bees.

Venom—the chemical injected into the skin when a honey bee stings. It's what makes being stung painful.

Virgin queen—an unmated queen.

VSH (Varroa Sensitive Hygiene)—Varroa Sensitive Hygiene honey bees can locate a larva in a capped cell that contains *Varroa*. VSH bees remove the infected larva and mites, and thus reduce *Varroa* populations in a honey bee colony.

Wax moth—larvae of the moth *Galleria mellonella*, which damages brood combs.

Worker bee—a female bee whose reproductive organs are undeveloped. Worker bees do all the work in the colony except for laying fertile eggs.

Worker comb—comb measuring about five cells to the inch (2.5 cm) in which workers are reared.

Resources

Books

The ABC & XYZ of Beekeeping, 41st Edition
Ed. by H. Shimanuki, Ann Harman
A.I. Root Co.

Africanized Honey Bees in the Americas
Dewey Caron
A.I. Root Co.

Aromatherapy Creams and Lotions
Donna Maria
Storey Books

The Backyard Beekeeper's Honey Handbook
Kim Flottum
Quarry Books

The Bee Book
Beekeeping in the Warmer Areas of Australia
Warhust & Goebel
DPI, Queensland

Bee-sentials. A Field Guide
Larry Connor
Wicwas Press

The Beekeeper's Garden
Hooper & Taylor
A&C Black Plc, London

The Beekeeper's Handbook
Sammataro & Avitabile
Cornell University Press

Beekeeping for Dummies
Howland Blackiston
Hungry Minds Press

Bee Sex Essentials
Larry Connor
Wicwas Press

Better Beekeeping
Kim Flottum
Quarry Books

The Buzz about Bees
Jürgen Tautz, H.R. Heilmann, and D.C. Sandeman
Springer

Candle Makers' Companion
Betty Oppenheimer
Storey Books

Control of Varroa for New Zealand Beekeepers
Goodwin and van Eaton
New Zealand Ministry of Agriculture

Covered in Honey, Cookbook
Mani Neall
Rodale Press

Dance Language of the Bees
Karl von Frisch
Harvard University Press

An Eyewitness Account of Early American Beekeeping
A.I. Root
A.I. Root Co.

Fat Bees, Skinny Bees
Doug Somerville
Rural Industries Research and Development Corporation

Following the Bloom
Douglas Whynott
G.P. Putnam

The Forgotten Pollinators
Buchman and Nabor
Island Press

Form and Function in the Honey Bee
Lesley Goodman
IBRA

Fruitless Fall
Rowan Jacobsen
Bloomsbury USA

The Hive and the Honey Bee
Ed. By Joe Graham
Dadant & Sons

The Honey Bee Inside Out
Celia F. Davis
Bee Craft Ltd

Honey Bee Removal
Cindy Bee
Root Publishing

The Honey Connoisseur
Marchese & Flottum
Black Dog and Leventhal

Honey Plants of North America
(reprint of 1926 Edition)
Harvey Lovell
A.I. Root Co.

Honey the Gourmet Medicine
Joe Traynor
Kovak Books

The Honey Spinner
Grace Pundyk
Pier 9

Increase Essentials
Lawrence John Connor
Wicwas Press

The Joys of Beekeeping
Richard Taylor
Linden Press

Langstroth's Hive and the Honey Bee
(reprint. 4th Edition)
L.L. Langstroth
Dover Books

Making Wild Wines and Meads
Vargas and Guilling
Storey Books

Natural Beekeeping
Ross Conrad
Chelsea Green Publishing

Observation Hives
Webster and Caron
A.I. Root Co.

Plants and Honey Bees
David Aston and Sally Bucknall
Northern Bee Books

Practical Beekeeping in New Zealand
Andrew Matheson
GP Publications, Wellington

Queen Rearing and Bee Breeding
Laidlaw and Page
Wicwas Press

The Sacred Bee
Hilda Ransom
Dover Books

Soap Makers' Companion
Susan Cavitch
Storey Books

A Spring Without Bees
Michael Schacker
The Lyons Press

The Superorganism
Bert Hölldobler and Edward Wilson
W.W. Norton & Co.

*The Thinking Beekeeper: A Guide to Natural
Beekeeping in Top Bar Hives*
Christy Hemenway
New Society Publishers

Understanding Bee Anatomy: A Full-Colour Guide
Ian Stell
The Catford Press

Weeds of the Northeast
Uva, Neal, and DiTomaso
Cornell University Press

Weeds of the West
Ed. by University of Wyoming
Western Society of Weed Science

Where Have All the Flowers Gone?
Charles Flower
Papadakis Publishers

The Wisdom of the Hive
Thomas D. Seeley
Harvard University Press

The World History of Beekeeping and Honey Hunting
Eva Crane
Routledge

Magazines

North America
American Bee Journal
Dadant & Sons
www.dadant.com

*Bee Culture, The Magazine of
American Beekeeping*
A.I. Root Co.
www.BeeCulture.com
*This webpage is the source of information for beekeeping
associations and state inspection agencies.*

Hive Lights
Canadian Honey Council
www.honeycouncil.ca

Australia and New Zealand
The Australasian Beekeeper
www.penders.net.au

The New Zealand Beekeeper
www.nba.org.nz

Europe
An Beachaire
The Irish Beekeeper
www.irishbeekeeping.ie

Bee Craft (UK)
www.bee-craft.com

The Beekeepers' Quarterly (UK)
Northern Bee Books
Jeremy@recordermail.demon.co.uk
www.GroovyCart.co.uk/beebooks

*International Bee Research Association
Bee World, and Journal of Apicultural
Research* (UK)
www.IBRA.org.uk

Norges Birokteren (Norway)
www.norges-birokterlag.no

Redaktion Deutsches Bienen-Journal (Germany)
www.bienenjournal.de/bienen/

La Sante De L'Abeille (France)
Federation Nationale
Des Organisations Sanitaires Apicoles
Departementales
Phone: 33.0.4.92.77.75.72
www.sante-de-labeille.com

Vida Apicola (Spain)
www.vidaapicola.com

Suppliers

B&B Honey Farm (general supplies)
5917 Hop Hollow Rd.
Houston, MN 55943 USA
Phone: 507.896.3955
www.bbhoneyfarm.com

BeeOlogy
Candles, soap, lotions, potions, and creams, plus much more
www.Beeology.com

Mann Lake, Ltd. (general supplies)
501 S. First Street
Hackensack, MN 56452-2001 USA
Phone: 800.880.7694
www.mannlakeltd.com

Ross Rounds
P.O. Box 11583
Albany, NY 12211 USA
Phone: 518.370.4989
www.rossrounds.com

Dadant & Sons (general supplies)
51 S. 2nd St.
Hamilton, IL 62341 USA
Phone: 217.847.3324
www.dadant.com

Betterbee, Inc. (general supplies)
8 Meader Road
Greenwich, NY 12834 USA
Phone: 800.632.3379
www.betterbee.com

Brushy Mountain Bee Farm (general supplies)
610 Bethany Church Road
Moravian Falls, NC 28654 USA
Phone: 800.233.7929
www.brushymountainbeefarm.com

Kelley's Bee Supply (general supplies)
807 W. Main Street
Clarkson, KY 42726 USA
Phone: 270.242.2012
www.kelleybees.com

E.H. Thorne, Ltd. (general supplies)
Beehive Works, Wragby
Market Rasen LN8 DLA
United Kingdom
www.thorne.co.uk

Pender's Beekeeping Supplies (general supplies)
28 Munibung Road
Cardiff NSW 2285
Australia
www.penders.net.au

Rossman Apiaries
P.O. Box 909
Moultrie, GA 31776 USA
Phone: 800.333.7677
www.GaBees.com

Photographer Credits

All photographs by Kim Flottum with the exception of the following pages:

Courtesy of A. I. Root Company/www.BeeCulture.com, 10; 11; 13; 61; 62; 134; 142 (middle, left); 143 (top)

Robin Bath, 162

Courtesy of Brushy Mountain, 142 (top, right)

R. Chamberlin, 57; 82

Ellen Harnish, 150

iStockphoto.com, 15

E. R. Jaycox, 8; 67; 76; 113; 114; 116

Walter T. Kelley, 142 (bottom, left & right)

Serge Labesque, 129 (left)

Courtesy of Mann Lake, 142 (top, left)

Allan Penn, 161; 163

Gwen Rosenburg, 9 (top); 22 (top & middle)

Courtesy of Shaw Traps, 131

Peter Sieling, 33; 34

Courtesy of Dr. James Tew, 58

Courtesy of United States Department of Agriculture, 53; 77 (top, right); 117 (left); 120 (top, left & right)

⇢About the Author⇠

Photo: Kathy Summers

KIM FLOTTUM brings a background of twelve years of plant science, honey bee research, and basic farming to his thirty years as the editor of *Bee Culture* magazine where his main occupation is finding the answers to the multitude of questions that beginning, intermediate, and even advanced and experienced beekeepers bring to the table. He teaches beginning and advanced beekeeping courses, travels extensively to educate and lecture, and contributes to a variety of other publications on the basics of honey bees and beekeeping biology, the business of bees and pollination, producing and using varietal honeys, and a host of other subjects. His books, magazine articles, interviews, and blogs are widely read for both their fundamental and advanced contribution to beekeeping knowledge. His magazine platform gives voice to his social commentary on topics ranging from genetically modified foods to pesticide abuse to both good and bad government regulations in the industry. He is beekeeping's leading advocate for fundamental honey bee safety including insuring excellent honey bee health, providing extraordinary forage, and minimizing the use of agricultural pesticides.

✦Acknowledgments✦

In case you aren't aware of this, there are only two beekeeping magazines published in the United States. *Only two*. The fact John Root chose me to carry on a tradition that was more than 120 years old when I joined his organization borders on the miraculous in my opinion (and is still considered so by some). And for more than twenty years I have had a voice in the United States, and to a degree, global beekeeping opinion. Who would have thought that an adopted kid from a small town in central Wisconsin would have that opportunity? Well, Chuck, Eric, and John seemed to have a feel for this . . . so now you know why I wake up every day with a thank you somewhere in mind.

Most certainly to my wife Kathy, my best friend, helper, cheerleader, and sounding board, who has put up with this craziness three times and helped each time. I couldn't do it without you. Thanks, Kath . . . more than you know.

And certainly to Bob Smith—friend, teacher, and without a doubt the stalwart beekeeping traditionalist in my life. I cannot imagine my life without Bob. If a person can be a father and a brother at the same time, Bob has been that person. And though gone now, you have benefited from his wisdom, his experience, and his humor, as have I. Thanks, Bob.

Of course this is also for Buzz and Nancy Riopelle—Friends for decades, helpers forever—especially Nancy for lotions and potions and candles and wax and common sense beyond measure and Buzz for everything beekeeping. Where would we all be without Buzz and the pictures I can take and the hints he shares? Thanks, Buzz.

And Chloe and Sophie for entertainment—they are cats, but cats are important late at night when deadlines are large, pressure dear, and the importance of the right things in life are in question. A purring cat offers a grounding that cannot be ignored.

And most importantly, to the thousands and thousands and thousands (and even more thousands) of beekeepers in my life who have shared what they have learned, what they know, and what they have done in the rest of the world.

And of course, the bees. Where would we be without the bees?

Isn't life grand?

Yes, it is.

→Index←

queen luxury, 167–168

queen mating, 168–169

queen productivity, 169

recordkeeping, 177

swarm management, 176

unproductive colonies, 176

Varroa control, 182–183, 183–184

Varroa research, 181–182

wax cleanliness, 173

winter season, 178–180

Russian bees, 54

S

safety

candles, 160

control strategies, 186

dangerous colonies, 186–187

health issues, 187

honey, 187

lifting, 185

robbing, 185–186

rules, 185–187

smokers, 92

stings, 185

wax, 153

seasons. *See* fall season; spring season; summer season; winter season.

Schell, Jeanne, 162, 164

Schell, Katie, 162, 164

scout bees, 60, 72, 74, 149. *See also* bees.

skeps, 13, 128

small hive beetles, 80, 101, 130–132

smokers

effect of, 44

cleaning, 92

emptying, 92

fuel, 42–43

lighting, 90–91

safety, 92

stings and, 71

solar wax melters, 156

spring season

American foulbrood treatments, 115

beeyards and, 20

broodnests and, 65, 146

capping in, 108

Carniolans and, 52

Caucasians and, 53

chalkbrood and, 112

colony division in, 149

comb replacement in, 107

drone-laying queens and, 63

essential oils and, 111

European foulbrood and, 113

food and, 24, 55, 83, 84, 99, 100, 102, 167, 169, 178

grease patties and, 117

hybrid bees and, 53

inspections in, 148

location and, 86, 170

packaged bees and, 48

Russian bees and, 54

swarming in, 148, 149–151, 176

tracheal mites and, 117

Varroa mites and, 117, 122

sticky boards, 120–122

stimulants, 110

stings

alarm pheromones and, 71, 72

allergic reactions to, 18

safety, 185

smoking, 71

stinger apparatus, 72

summer season

bee suits and, 41

beeyards and, 20

broodnests and, 65

Carniolans and, 52

chalkbrood and, 113

food and, 24, 83, 84, 178

hive insulation in, 85

honey harvesting, 38, 135, 138–144

Italian bees and, 50

location and, 86

Varroa mites and, 122, 126, 178, 182, 183

ventilation and, 29

wax moths and, 127, 128

water consumption in, 16, 24

supers

adding, 101

cut-comb frames, 133

small hive beetles and, 131–132

supersedure. *See also* queens.

aggression and, 51

chalkbrood and, 112, 113

emergency replacements, 61–62

European foulbrood and, 113

hive balance and, 62, 136

purchasing, 99, 149

queenless colonies, 60–61

recordkeeping, 103

spring season and, 149

swarming and, 54, 148

surplus honey, 135

Survivor Bees, 122

Survivor Queens, 54–55

swarming

catching, 149–151

causes of, 59–60

management rule, 176

queen cell cups and, 60

splitting colonies and, 148–149

supersedure and, 54, 148

urban beekeeping and, 23

T

temperament, 51

Tew, James E., 144

tools. *See* equipment.

top bar hives

comb inspection, 36

disease control, 40–41

follower boards, 36

frame examination, 39

harvesting, 37

installing, 36–37

introduction, 34–35

pest control, 40

winter and, 38, 40

tracheal mites (*Acarapis woodi*), 54, 116–118

tremble dance, 77

25 Rules For Modern Beekeeping, 188

U

uncapping, 152, 177, 187

urban beekeeping

bee density and, 24

☙ Also Available from Quarry Books ☙

Better Beekeeping
978-1-59253-652-8

The Beekeeper's Journal
978-1-59253-887-4

*The Backyard Beekeeper's
Honey Hanbook*
978-1-59253-474-6